D1631884

AUDITIONS

AUDITIONS

Architecture and Aurality

ROB STONE

THE MIT PRESS   Cambridge, Massachusetts   London, England

MIT Press books may be purchased at special quantity discounts for business or sales promotional use. For information, please email special_sales@mitpress.mit.edu.

This book was set in Minion and Meta by the MIT Press. Printed and bound in the United States of America.

Library of Congress Cataloging-in-Publication Data
Stone, Rob, 1962 August 31–.
Auditions : architecture and aurality / Rob Stone.
     pages   cm
Includes bibliographical references and index.
ISBN 978-0-262-02886-8 (hardcover : alk. paper)
1. Senses and sensation in architecture. 2. Sound in design. I. Title.
NA2543.S47S76 2015
720.1—dc23

                              2014034233

10   9   8   7   6   5   4   3   2   1

Contents

Acknowledgments

This book has taken up quite a bit of time and space in my life, and in those of others. So, there are some people I have to thank. First, Irit Rogoff, Jorella Andrews, Gavin Butt, Brendan Prendeville, Andy Lowe, Astrid Schmetterling, and other members of the Department of Visual Cultures at Goldsmiths College with whom I worked closely during the initial phases of thinking about this book. I'd also like to thank the Department of Art, Philosophy and Visual Cultures at Middlesex University for time and material assistance when pulling the book together.

Longstanding conversations with Barbara Engh and Peter Nix have identified not only substantial areas of interest, but also the ways to approach them. For setting me off in a general direction a long time ago, I would like to thank Gen Doy and Pat Kirkham, and Fred Orton, Griselda Pollock, and Ben Read. Thanks also to Rebecca Duclos for reminding me what I was doing at an important point. Roger Hewland, the affable owner of a second-hand record shop in south London, was endlessly helpful and patient with me, joining in researches that often led nowhere, as did staff of the libraries of the Royal Institute of British Architects, the Wellcome Institute, the Chicago Historical Society, the British Library Sound Archive, Fondation Le Corbusier, Senate House Library, and Goldsmiths Library. Thanks also to the artists, archivists, and architects who have generously allowed me to use images and who have commented on portions of the book—especially Eleanor Antin, Tadao Ando, Elizabeth Diller and Riccardo Scofidio, Guido Guidi, Joan La Barbara, Bernard Tschumi,

Neumann GmbH, Bayreuther Festspiele GmbH, and Deutsches Theatermuseum München. A period of research for the book was funded by a grant made by the Leverhulme Trust, for which I am extremely grateful.

There are some for whom simple thanks aren't enough. Roxy Walsh, Simon and Anton Ofield-Kerr, and Janice Cheddie have long had a profound influence on me. I am also greatly indebted to Adrian Rifkin. His guidance has been by turns Delphic and direct, and always liberating. This book is for them. It is for my parents too. And, it is for Heidi Reitmaier (who has given me a lot of slack in my more domestically agitated moments), for Gabriel, and for Ava.

Introduction: Architecture, Aurality?

By pleasure, I mean I notice what happens.
I can attend to it. Rather than, as you say,
surrender, I can pay attention and become
interested in what it is you are actually
interested: in what superimposes what?
—John Cage

Morton's Minority

"Call it culture," he said, speaking of the intrusions into our lives of the potentially upsetting activities of others. There is a technical term belonging to the hinterlands of music criticism. I wish it were mine; but it belongs to the vocabulary of that complicated musical voluptuary, Morton Feldman. The term is *uninfluential*, a word Feldman used both occasionally and anecdotally. By this I mean that the constancy in his usage of it was highly specific. The regularity, the stability of the uninfluential arose for Feldman not from any pregiven dictionary definition. Rather, it stemmed from his ongoing aggregation of expanded and diverging insights into what else it was that music might happen upon. He didn't adhere to a universal convention with the word, but his exercise of the term wasn't completely idiosyncratic, either. Even if they did not use the term themselves, peers and collaborators such as fellow composer John Cage in one sphere, filmmaker Robert Frank in another, could at least sympathize with the ambitions of the uninfluential as part of a modern minor diction.

Producing its own constancy, even constellating its own milieu in this way, the uninfluential might be seen as Feldman's intellectual emblem. Sometimes it appeared as a name for fugitive qualities to be sought *in* music, or sought *by* music. At other times he hinted with it at traces of nameless vernacularity in expired traditions. It could indicate some singular gravity in an admired artist—Anton Webern or Piet Mondrian, Philip Guston or Frank O'Hara, for example. It could be the object of a kind of aesthetic attention. Or, more importantly, it could be the form of that attention itself: concrete, yet virtuously unseizable. Feldman saw in the enigmatical condition of the uninfluential a possibility that was at once musical, intellectual, and social—and also a kind of relief, a kind of reward for living a life in a close relationship to an understanding of art as being already philosophy, already criticism, already some form of social theory.

———

For Feldman, the uninfluential was a product of the social character of one's engagement with a work of musical art. Certainly, he mentioned it as something to be looked for in the absorptions of shared listening. Observed through an effort of attention to the effects of superimposition or intrusion, appearances of the uninfluential might then also outline a type of sociable compact and become a thing that one may wish to converse upon, invest in, or clarify in their fleeting state. To enable that speculative dialogue, however, one might also need to produce grounds for the recognition and reception of such newfound objects and qualities; that is to say, one might have to create the grounds for their entry into a shared and pertinent language, no matter how secluded, fragile, or densely allusive that language might need to be.

Feldman wasn't alone in insisting that such creative-cognitive demands be the outcome of new experiments in sociable listening. Neither was he the only one to propose sociable listening as itself the start of a reimagined musical aesthetics. Cage probed similar terrains of attentional poetics. So too, in their own ways, did Christian Wolff, Earle Brown, and others of the New York School in the 1950s and 1960s. In doing so, they joined with a growing international community of experimental musicians in the pursuit of traces of novel social and aesthetic being. These were traces that, in their necessary modesty, in their minority, were so radically lacking in grander significance and portability that they could scarcely be noted, let alone misrepresented or exploited by figures who seemed to speak for new and bullying hegemonies in modern music. For Feldman, the ungovernable and creative qualities of anecdotal listening that were signaled by the concept of the uninfluential were what prevented one from being eaten up by the eagerly co-opting economies of a villainous Pierre Boulez or an inauthentic Karlheinz Stockhausen.

———

At times, Feldman was increasingly less committed to the principles of fortune or improvisation as motors of musical composition and performance than were his immediate musical associates. Or, more correctly, his interest in chance and invention was focused elsewhere in the apparatus of musical meaning. He controlled the aural materials that went into a space and appeared under his name in a direct, assertive manner. Rhythms and timbres, durations and pitches, attacks and decays, intervals, chords and clusters, rests and ornaments and progressions; all these things were subject to his intense discernment and specification. However, his authorship of the listening dimension of his music was almost opposite to that. As far as he was concerned, one needn't feel obliged to sit in obedient silence, or even hear the entire performance of a late, long work such as *For Philip Guston*. Rather, a person might sit or stand or promenade. One might find oneself moving, unassertively, though not passively, into and out of the social sphere produced by the textures of the music. In so doing, a person might discover an interest in an entirely different work of art, in a view, in an other. One might leave, return, be riveted or suddenly lost. One might reflect, converse, remark upon details of the architecture produced by the sound, its extension, the character of its duration, and whatever else it occasions. In the space made by the music, rather than over or against it, one might discourse, find or strike up aesthetic and other kinds of relationships—one might start to recognize, in short, the space-making capacity of the sound and the condition of one's public, intellectual self in it.

Feldman wasn't an architect. He was interested in rugs and Mies van der Rohe, it is true, but he was interested in a lot of things. However, he was also able to pursue a recognizable social, even civic philosophy through making modern music and talking about how one might attend to it. He liked the idea that those who attend to artworks are the only ones in any kind of position to complete them. He liked the fact that there is no policing this, and he took seriously the notion that whatever it was that might come about through this completion, it was that and only that which could be understood to be intended by the work.

By placing acoustic materials in such a way that they became architecture themselves, while at the same time transforming the existing architecture of a place and, in the process, facilitating experimental, speculative modes of association, Feldman made an inhabitable relationship between sound and space a characteristic of musical modernity. With that convening gesture, he also advocated a form of creative attention appropriate to such spaces. He gave voice to a type of attention that, rather than pretending at passive objectivity, admits to the way it sets about constituting artifacts for itself—taking the available musical and spatial circumstances as a start and supplying a kind of counterpoint by drawing in a thoughtful metonym from here, a fragment or relic from there and other kinds of part-object from elsewhere to manufacture the kinds of intrusions, super-impositions, afterimages, intersections that are prompted by the structure of his music. What can be found in Feldman's works, then, is both a medium for subjective acts of assemblage and a theory of association. As such, they are as theoretical, and as compelling, as similar theories articulated by figures as dissimilar as Richard Payne Knight and Sigmund Freud. Produced in the context of an enormous liberalization in understandings of the social capacities of art while at the same time remaining in the grip of Feldman's own curiously plangent machismo, the works' precise character and progressiveness will possibly remain unsettled.

Familiar Passages

One thing this small sketch of Feldman might prompt is a suggestion that the notion of aurality with which this book is concerned has something to do with creative attention and also something to do with the potential for new civilities. It isn't a book about the politics of listening to experimental music, however; not quite. It is a book about architecture, and some aspects of the frequently rather Delphic relationships that have arisen between modern sound and modern architecture in the period since the mid-1930s. It aims to take some steps toward apprehending architectural sound culture in a way Feldman might recognize and to start to understand it socially and in its legible, concrete, and perceptual complexity.

I should make some brief preliminary remarks. First, this field of interest—it can't be called a discipline—is almost oceanic in scale and diversity; it can scarcely be said to have a unifying literature, and much less can it lay claim to a group of canonical objects or exponents or a set of methodological approaches. Second, it isn't new. Architectural and musical discourses have long been speaking about themselves in each other's terms. "Frozen music" is a familiar simile, for instance. It is difficult to imagine an appreciation of a Bruckner symphony that doesn't nod to a view of the music's architectural qualities. And, just as an example of one of those responsible for drafting some of the founding documents in the professionalization of architectural history as a discipline in the early 1940s, Rudolf Wittkower was keen to argue for the difficulty of understanding the architecture of the Renaissance and Baroque without a grasp of the theories of proportion and harmony that were being carried on in the music of the time.

Third, I've chosen the mid-1930s for metaphorical effect here, rather than as an indication of the start of a period to be chronologically accounted for in a developmental manner. At this time, the way people talked about architecture started to change dramatically. Already in the 1920s, figures like Walter Gropius and Le Corbusier, with their polemical buildings and writings, had begun to refigure the expected responsibilities, look, and organization of architecture. Though they hardly represented a fraction of the changes in building culture at the time, these architects started to change what was expected of the word architecture. And, by the mid-1930s, historians as contrasting as Nikolaus Pevsner and Henry-Russell Hitchcock had set about rewriting histories of architecture in order to account for its modern appearance and the new social and aesthetic concepts through which architecture seemed to be coming to define itself.

To a certain extent, the relationships between modern architecture of the kinds that interested Pevsner and Hitchcock and, for instance, Anton Webern's musical exploitation of the twelve-note row or the particular emergence of Duke Ellington or Louis Armstrong's jazz is already familiar enough. But, at this time, music wasn't really the predominating aesthetic force in the acoustic reconsideration of architecture and the ways that it extends and endures. The 1930s also saw dramatic acoustic changes. Telephony, phonography, public address systems, broadcast radio, synchronized film sound, issues like increased noise from road traffic or typewriters, the unnerving quietness of suburbs, even the notion of a "talking cure"—all powerfully altered the sensible fabric of spatial modernity. And, of course, these novelties didn't stop appearing. The development of sonar and other listening technologies during the Second World War fed directly into the music industry afterward. The new spatialities subsequently afforded by the availability of stereo recordings came to foster a contest for commercially unique acoustic signatures among recording companies in the postwar period. With this, sound became inhabitable, and educated choices could be made about its sources and how to savor it.

With this technical realization of the inhabitability of sound, it is also possible to imagine processes of its commodification that operated between, on one hand, the particular penchants for domestic acoustic intimacies, an almost libidinized set of attachments to the details of certain brands of commercial music, and, on the other hand, the generalizing desire at the time of an organization like the British Broadcasting Corporation to reach through into that private realm to produce a national community via a program of cultural education. In this way, it is possible to see a nexus of cultural ideologies, technologies, psychological processes, and economies start to animate the idea of sound as livable space. At the same time it is possible to see the origins, perhaps, of the agitation in Morton Feldman's tone when writing about the necessity and the difficulty of sustaining a more than individual relationship to music.

As a spatial discourse at this time, then, as a spatial practice, not only did architecture come to exist in relation to these other, acoustic spatialities, but, in their inhabitable state, these acoustic formations, by turns utopian and quotidian, came to take on the properties of architecture itself. In the mid-1930s it is possible to see new kinds of cultural artifacts start to appear that are both spatially complex and at the same time insufficiently explained by the poetics of existing architectural discourses or existing musical discourses alone. This perplexing insufficiency for the decoding of modern cultural change didn't pass by unremarked upon even at the time. Rudolf Arnheim's 1936 book *Radio: An Art of Sound* and Theodor Adorno's early work on the phonograph and radio from the same time are marked both by necessary acts of imagination of sound's capacity to produce concrete spaces and accounts of how states of modern subjective being come about in those spaces.

Arnheim applauded mildly. Adorno wasn't at all convinced about the situation. But, notably, both these writers (there were many others) recognized that phenomenal aspects of what they were becoming interested in were appearing in unprecedented ways and through effects of intrusion, superimposition, and the like that Feldman described. Arnheim became very excited by the narrative possibilities of consonances between a snore and the puffing of a train. Adorno characterized intellectual space by remarking on the sound of his slippers as he shuffled between his desk and his phonograph. He castigated bourgeois taste by rhyming it with the hoarse, spittling sound of a businessman's soda syphon. The experience of hearing someone whistling a snatch of Brahms at a train station prompted in him a heated evisceration of a recording industry that was capable of evacuating the philosophy from Brahms's music and reducing it to a ditty, all in the name of higher culture. The issue at stake in the appearance of these anecdotal listenings is that they have a powerful, almost totemic function in the oeuvres of their authors. They convoke unevenly advanced areas of inquiry while announcing a point of their intersection as the moment and the structure of an otherwise impossible critical observation on the world. The history of this anecdotal aural practice has since involved some of the most significant critical gestures made by figures as diverse as Adrian Stokes, Julia Kristeva, James Baldwin, Roland Barthes, and others working with different kinds of intention.

Ariadne's *Ought*

It is possible that I remember the following event with rather too much clarity for it to be entirely plausible as an actual occurrence. But I recall an occasion in the mid-1980s when I was studying the history of architecture in an undergraduate course. I was looking at a set of drawings made by Le Corbusier. I remember the details of the architecture of the library I was in, the color and pattern of the carpets, the etched grubbiness of the surfaces of the desks. The images I was looking at, simple, seemingly so at least, were of those small, familiar buildings typically figured as the heroic modernist riposte to crises in the European housing stock made by the rhetoricians of the Congres Internationaux d'Architecture Moderne.

I remember looking. And, I remember thinking about the convincingness of the expanses of white, un-marked ground between the buildings. Over these expanses, as a kind of signature, romped the tiny Virgilian characters with which Le Corbusier gestured toward the population of his architecturally ideal terrains. Too early at that point in a particular process of art historical instruction, I'd yet to suspect the implications of white ground as a type of expectant political figure. The liberties that its formulation and reformulation might afford modernist art practices hadn't yet occurred to me. Neither had the constraints they might imperceptibly place. But I do remember thinking about how these images should sound.

Should I hear a timeless wind? The sea and gulls? Or should more determinedly cultural, specifically French rustications resonate? Gluck's flute, or Debussy's? Duparc's *melodies*? Poulenc? Poulenc, possibly. Though for that more urban register perhaps the lilting of Lucienne Boyer's voice, or some of the North African–inflected jazz that was favored in 1930s Paris. The jouissant euphony of the gamboling figures Le Corbusier depicted, the hectoring gravel of his own accent, the sporty honk of a distant Voisin motor, even

I.1 Le Corbusier, *Réorganisation agraire, ferme et village radieux*, from *La Ville Radieuse*, 1938. Plan FLC 28619, © FLC/ DACS.

Le Corbusier's later musical experiments with composers Edgard Varèse and Iannis Xenakis; how reliably *ought* these things behave in an interpretation of such dreamed architectural geography?

It wasn't long before I forgot any and all iconological probity and started happily substituting songs by the Clash or the distinctive idiolect of the much loved travel journalist Alan Wicker just to imagine what then happens to that architecture. In a cinematic sort of way, ventriloquizing Le Corbusier's drawings by ringing changes in their soundtracks revealed an entertainingly transformative narrative capacity. I realized I already knew this as a popular procedure of creative defamiliarization from watching those precursors to MTV music videos on late-night BBC music programs like *The Old Grey Whistle Test* in the 1970s. Le Corbusier's drawings weren't just the comical dupes of a faux surrealist technique here. They showed a type of conversational resistance and, as *plans*, his drawings also suddenly appeared self-confidently as *scores*. It occurred to me that while these drawings represented a kind of solfège for architectural pedagogy, the musical and other aural anachronisms I was playing with suggested that they could exist as a kind of musical graphic notation at the same time. Those architectural figures, the white ground and those rudely playful athletes, might legitimately represent some portal to an understanding of architecture's improvisable relationships with a gamut of aural modernities.

The temporally commanding *ought* that was caught up in these wonderings struck me as somehow correct, yet incongruous, even then. There was some uneasiness for me as, at that time, a student of a particular kind of art history who had a special commitment to contemporaneity as the basis of the reconstruction of historical meanings for cultural objects. Looking back, I think I was searching for some kind of justifying evidence of uncertainty in that method; evidence that could allow a degree of freedom and collateral speculation about the nature of audiences to artworks at the time of their making.

Ever a diligent and well-behaved student, as it happens, at that time I was also attempting to give myself an informal education in opera. I now know more about opera than I did then, but I have never been able to appreciate it on what I take to be its own grounds. I'm interested in it, but I don't *like* it. That said, I can listen for hours to Anne Sofie von Otter trying to sing the opening syllable of Handel's *Ombra mai fu*. Listening to the way that she moves the sound around her buccal architecture, the way she climbs about to change the necessary shape of the vowel for musical effect, introduces one to an unlikely and intimate spatiality as the subject of listening. It is a space that has less to do with the dramas of stereo microphony than it has to do with the techniques devised by a singer whose voice is musically accommodating the physical and artistic changes that come with age and improved perception. As such, there is a touching quality there, a poignancy that is only heightened by the comparison of her voice with the uncompromising, pneumatic certainty of, for example, Bryn Terfel singing the same notes—a singer whose sentimental responsibilities to the manly, industrial landscape of the Welsh bass-baritone are perhaps rather different from the ones that von Otter's voice articulates. What I am trying to indicate here is that it is the eccentric, unlikely collateral effects of her voice that are interesting, not necessarily what I *ought* to be measuring as strictly operatic competences. This is perhaps a form of superimposition.

The particular piece I was listening to at the time I was caught in a study of Le Corbusier's little drawings was a recording made in the late 1960s of Richard Strauss's "opera within an opera" *Ariadne auf Naxos*. The production involved Rudolf Kempe conducting the Dresden Staatskapelle. Gundula Janowitz sang Ariadne and Prima Donna. Sylvia Geszty was Zerbinetta. The performance was recorded by EMI in the Lukaskirche in Dresden, a venue known for a particular, even recognizable reverberance, and a place we will be returning to later. I had borrowed this recording from a library, copied it to tape and was listening to it on a Sony Walkman—on the bus, on my bike, just trying to get used to what was for me a strange new musical environment. The quality of the recording wasn't good. There was no libretto, no sleeve notes. Not only did I not understand what was being sung, not only did I not know the names of the singers, still less their reputations, it wouldn't have mattered much if I did. I scarcely knew who Ariadne was, and I definitely wasn't aware of the way Strauss had so frequently used such mythological women allegorically to explore his own fraught relationships to the culture of modern, industrial Germany. What I *was* aware of was that, as Gundula Janowitz sang her way through Ariadne's aria "Es gibt ein Reich," it became clear to me that she was singing the building. Literally an intuitive apprehension, it seemed as clear to me as it also seemed unlikely that there, in the way her voice played to the resonances of the Lukaskirche, was a meaningful operatic object, musical and architectural. Her voice animated the architecture of the Lukaskirche in ways not unlike the ways that, later, the soprano Joan La Barbara was to explore Richard Serra's cavernous steel sculptures. La Barbara's explorative, evocative soprano voice is burdened by a narrative of her relationship to Morton Feldman and other figures in American experimental music. Janowitz's, at this moment, is burdened otherwise.

The apparent fact that Janowitz didn't sing over and obliterate her own resonance, but rather let the building sing, I later found had its own significances that were internal to the opera. Ariadne and Prima Donna are roles sung by the same singer and are echoes of each other. Because of what goes on before Ariadne's aria, where words and melodies are sung back by Echo to those watching over her, and the fact that Ariadne finds herself alone in a shattered existence as she starts to sing, it seems to make perfect operatic (if not exactly narrative) sense that Janowitz should at that point bring the voice of the building itself onstage. Maybe this is just a lightly touching narrative effect—simply the image of a lost mythological soul finding solace and company in imagining the reflections of her own voice as being that of another. But, perhaps there is more implied in this space, a kind of acoustic mirror—a sense of the building being suddenly considered the source of a Gaze that lets Ariadne know that, although she is as far as she is aware alone, she can still know herself as being watched, if only by herself. If Strauss didn't intend *Ariadne auf Naxos* as a complex allegory of the spaces made by the recorded female voice when he wrote it in 1916, it certainly can become that.

When I suggest, then, that aural-architectural space may produce complicated subjects and objects of interest, it can be of this order of complication—made by drawing in concatenations of fiction and representation and attention and psychoanalysis and mythology and microphony and engineering and tradition, and so on. It is not possible to be certain that Janowitz was intentionally singing the Lukaskirche as an operatic instrument and extension of her own vocality. Perhaps, because of this, the ontology of such aural representations of architecture should be regarded as especially fugitive, uninfluential, while bearing enough plausibility to be worth beginning a speculative conversation about.

Moving Ariadne's aria to different venues produces different phenomenal and interpretative effects. If we take the historical recordings of it made by sopranos Lotte Lehmann or Maria Jeritza and substitute these as soundtracks to Le Corbusier's drawings, different issues seem to manifest from the architectures that are respectively captured by each form of depiction. By bringing these particular recordings to Le Corbusier's drawings, one might note that modernist buildings of the 1920s and 1930s, in some visually discursive way, manage to retain at least the rhetoric of their modernity, whether sketched or photographed. The same can't be said of sound recordings made at the same time. The acoustic sepia in these recordings might signal, for instance, the fact that architectural time exists differently in phonographic and photographic form. More, one might note that a privilege is given to photographic representations of the textures, volumes, and ornaments of architectural space. While these things are indeed captured by phonographic representations of architecture, they seem to remain enigmatic. A photographic architectural sign is not the same as a phonographic architectural sign. This is almost self-evident. So, one of the aims of this book is to try to provide vocabulary for graphical effects similar to that of Janowitz's animating voice that keep intact the enigmatic spatial condition of what they happen upon.

Photographs and Phonographs

The work of an agenda-setting acoustical pioneer like Leo Beranek in the early 1960s remains important to anyone broaching the field of architecture and aurality.[1] Producing detailed visual and aural images of concert halls in Europe and America by photographing and documenting his measurements of rates of acoustic decay, Beranek's impressive empiricism stands in the long, respectable tradition of approaches to aural-spatial perception. It is a tradition founded in the nineteenth century by physiologist Hermann von Helmholz.[2]

Helmholz's concerns with the minutiae of harmonic partials have come to profoundly influence the study of the physical mechanics of music production in terms of architecture and in terms of voices, instruments, and otherwise. Specific schools of thought about the psychoacoustics of musical perception have been sponsored by Helmholz's work, which has also determined the approaches taken in practical manuals produced for the architectural profession regarding the suppression of pesky eigentones or neighborly noise.

The opera house isn't the most important of aural-architectural apparatuses in Western culture, but it is one of them. This is especially so when one considers the very dense history of staging and restaging operas, changing tastes in the preferred design of buildings, and changes in the character of different schools of singing over time. However, Beranek's Helmholzian attention to the acoustics of opera houses might now be viewed as being if not misplaced, then in some ways peripheral when it comes to such cultural concerns. It is worth noting that when Jean Nouvel provided a plan for concert hall in Lucerne in the late 1990s, the acoustics of the hall, though much praised, were not of primary importance in the way the building was talked about.[3] Rather, Nouvel's sentimental attachments to the opera house as a rich seam of bourgeois architectural culture were phrased in terms of the civic and professional embarrassment that could ensue should he mishandle the view of the city from the foyers. As an extended pedagogical, aesthetic, municipal, and architectural enterprise, the dignification of the sociability of a particular class is an unavoidable aspect of opera's achievement. In this sense, as the sound (by now we can call it the space) made by the bourgeoisie, opera takes on an intriguingly onomatopoeic value.

———

If Helmholzian physics is thwarted by its desire to explain rather than further poeticize these spaces, there is still another resource. In 1977, Michael Scott brought forward the first volume of *The Record of Singing*—a multidisk volume released by EMI.[4] The stated aim of this survey of early phonographic recordings was to provide an introduction to different national schools of operatic singing at different moments. With its subsequent volumes, the project covered the period from 1899 to 1952, stopping short of the commercial advent of stereo. The handbooks that Scott wrote to accompany the series remark on voices, performances, and fragments of biographical material. They make clear but do not examine in qualitative detail the relationships between singing and changes in recording technology. Despite that lack of an analysis, *The Record of Singing* remains a publication that makes available a canon of microphonically realized architectural spaces. As much as different styles of training are heard to evince in the affective qualities of the voices of Claudia Muzio or Titta Ruffo, so too are the characteristic acoustic blooms of opera houses in Milan, Munich, and elsewhere. At the same time, Jürgen Grundheber produced on vinyl a performance history of the Berlin Lindenoper. While archivally indicating a local cultural history of favored repertoire and performers, Grundheber also demonstrated the significance of historically specific microphonies at work over time in the representation of the same building.[5]

As further examples of what else music might happen upon, these two histories are important architectural documents. They are important for the remarks made and the performances collated, but they are also important because they have the capacity to open out onto a much broader view of ways of accessing aural-architectural space—ways that are not in and of themselves acoustic, and which might play with a notion of the architectural drawing as a *score* rather than as a *plan*.

Mary Ann Doane has shown phonographic space to be the subject of a complex physical and subjective acoustics. In particular, she has put in place the ideological concerns that bear on the perspectival framing of the female voice in cinematic and other forms of recording.[6] One thing that is clear from her work is the instrumentality of the recording studio in producing this space and the way it appears. Contemporary multimiking techniques and the use of convolution software to mimic the acoustics of known concert halls are just a recent turn in a history of changing spatial aesthetics, recording practices, and technologies. There is a history of auteurs in the field. Arthur Haddy, the recording engineer at Decca, could be one for his part in the development of the Decca Tree microphone array. Walter Legge could be another. His engineering of the different environments conjured by Herbert von Karajan's conducting of Rachmaninov's orchestral works in the early 1960s produced one of the most pervasive and recognizable of commercially available sonorities. Arthur Lilley, with his glossy honing of Mantovani's arrangements of *Charmaine*, could be another. These were practitioners and inventors of musical confection with myriad tricks of the Foley artist in their repertoire—placing this type of microphone and not that, here and not there, opening a window, mixing different ambient and instrumental sources together, editing, and so forth. While EMI and Decca and others commercially refined these stereophonic spatial nuances, Westminster Records continued well into the 1960s to work with the renaissance luminosity attainable by the just placement of a single microphone to record a performance. At the same time, the same industry produced the baroque spatial distortions exploited to popular acclaim by the likes of Phil Spector and Brian Wilson, to say nothing of the work Daphne Oram, Delia Derbyshire, and Pauline Oliveros were also doing.

To think of Legge, Lilley, and Haddy as spatial artists as refined in their studio practices as architectural photographers such as F. R. Yerbury, Cervin Robinson, or Julius Schulman reveals productive tensions between what phonography and photography are able to spatially provoke, apprehend, and anchor. The photographic deportment of performers on record sleeves and in publicity materials, the ways they conduct themselves—how von Karajan wore his nails and hair or his clothes, or how Mantovani folded space with his hands or simply stood—are issues of visual representation that often appear to flatly contradict the spatial fictions regarding the place and moment of performance evoked by commercial recording techniques. The industrial musical culture inhabited by the likes of Haddy, Legge, and Lilley was an advanced one that was consciously involved in making complicated and variably plausible spatial fictions. These contradictions between the phonographic aural fictions that appeared on vinyl and the photographic aural fictions that appeared on record sleeves and other promotional materials also seem to prompt a distinction between notions of *studio* recording and *field* recording.

————

During the 1930s in the United States, the Resettlement Administration and, later, the Farm Security Administration engendered a rather new ethics of recording. Here the work, methods, and intentions of the champions of an American musical ethnography—musicologists like Charles Seeger or John Lomax—are important in the way they came to be, in a kind of documentary fashion, haunted by claims of indexical authenticity. Seeger's anthropologically ethical appeal to his own field workers to "record everything" seems to signal an objective lack of discrimination and a desire merely to archive whatever traces of an evaporating traditional culture could be found. Through that lack of discrimination came about the disparate expansiveness of the Archive of American Folk Song. Seeger's views on the role of all successful vernacular musics in welding a "community into more resourceful action for a better life" resonates with sentiments articulated by photographers of the same subjects at that time, such as Arthur Rothstein and Dorothea Lange.[7]

What I mean here is that, for the idea of field research, it is important to understand the respective roles of the Presto sound recorder used by Seeger, Lomax, et al. and the Big Bertha camera used by Rothstein, Lange,

I.2   Arthur Rothstein, *Cabin*, 1938. Library of
      Congress (LC-USF34-TOI-8859-D).

et al. in the making of a portrait of the anonymously American, rural vernacular culture that was pursued under the auspices of the Resettlement Administration. The continuities and discontinuities allowed in the relationship between the two media might hint at the tensions between well-intended documentary realisms and the elegia that came to form around the photographs and the phonographs, and which later would become commodified by a music industry keen to issuing and reissuing blues and Cajun music as part of a support structure for a further categorization of different types of marketable urban music genres.

—————

With these examples, what I have been attempting to introduce is a field of aural-architectural attention and just some of the questions that follow from the character of that attention. It is quite possible to imagine a set of plausible connections between the cabin home of an African-American family in Alabama as photographed by Arthur Rothstein and the drawings by Le Corbusier that we started with. We know Le Corbusier had interests in vernacular architectural types found in North Africa—something that went hand in hand with the close sense of ownership that French colonial ideology imbued in its subjects, no doubt. We know that in the 1930s he worked in a global context of a desire for state-level responses to shattered lives and shattered economies, and that from that response grew powerful visual sources of national cultural ideology. This might help join his drawings to Rothstein's photographs too. That same conversation might also have good grounds to connect the Great Depression to the elegant shack Le Corbusier built to remind himself that he was still in a position to keep an eye on Eileen Gray and her domestic arrangements at E-1027.[8] However, it is perhaps clear now that the possibility also exists to connect these dreamed buildings differently via a series of auralities—to the extent that the cultural history and spatial modernity of a guitar lick by Keith Richards or Mick Jones might eventually sit so much more plausibly between them.

1.1   John Vachon, *Unemployed Men Who Ride the Freight Trains from Omaha to Kansas City and St. Louis and Back*, November 1938. © Library of Congress (LC-USF34–008859-D).

1 Ballerinas Are Always Hungry

There is only conservation of that which has
duration—of that *on which time gnaws*.
—Gabriel Marcel[1]

History is perverse, random, and mercenary.
—Eleanor Antin[2]

Here we consider pasts present: old sounds, old architectures, and the elegiac architecture of informal conversa-
tion. Observing older thresholds of meeting and departure, Fletcher Moss and Charlie Chaplin show garden gates
as the furniture of civil encounters. In developing a concept of aural dust from this, Eleanor Antin and Gabriel
Marcel suggest how these things segue socially into Oscar Newman's "acoustic window" and the theology of new
urbanism.

19

1 I Thort Your Kindness Was Love, but It Ain't

In 1915, at the garden gate to his home, the Old Parsonage in Didsbury, just outside Manchester, Fletcher Moss stood at a threshold in more senses than one. A civic-minded soul, he knew this. As a local architectural historian of merit, Moss spent the first twenty years of the twentieth century publishing a lengthy series of pilgrimages to old homes in Cheshire and other counties along the Welsh border. These companionable episodic travelogues contain brief political and religious histories of the Recusancy in the older houses about Cheshire. The books also include examples of James Watts's architectural photography, some of the most elegiac committed to print. Not exactly prone to nostalgia, Moss was nevertheless interested in somehow making the past present.

Not only this: in his municipal capacity as Alderman Moss, he also quietly influenced the thinking of those younger political cadres that were committed to the modernization of Manchester. By the mid-1930s his thinking had led, against fearsomely organized conservative opposition, to the implementation of Barry Parker's designs for Wythenshawe as a satellite garden settlement of Manchester, a new airport and town hall, new libraries in the city center and its suburbs, and the building of those suburbs themselves. There were also provisions for the regulation of new independent bus companies, a radical and progressive program of slum clearance and road building, and an eventually aborted proposal for an underground transport system.

In the same year, 1915, Moss privately published an autobiographical account of his involvement in the transformation of the small but viable community of Didsbury from an ancient hamlet with a population of little over a thousand, a recognizable and legally incorporated suburb of Manchester—one that was preferred by the city's liberal and cultured scions, with a population accelerating now past fifteen thousand.

His was no ordinary garden gate. Still standing today, a massive and surreally impressing affair, his gate was composed from remaining fragments of Manchester's demolished civic and commercial buildings.[3] Moss recalled an illuminating incident that took place there. A crisply modern, city-dressed gent approached him asking for directions to the town center. Pointedly, Moss said that as far as he could remember, this, where his gate stood at the garden to the Parsonage, was Didsbury's center; but if one is looking for the train station, it is in that direction (he gestured toward the Midland Line). Maybe, given the date, he sensed and resented the air of the officer class in the tone of the stranger, and dismissed him accordingly. In any case, for Moss, the droll triviality of this informal instant of accidental public association had significance as a parable. It indicated a substantial geographical reorientation of sensibility wherein the Parsonage had become merely quaint in its cultural significance, old and no longer a vital social and spiritual hub. The meeting metonymically implied Moss's grudging recognition both of modern Manchester as the distant center and his own administrative culpability in the dormitorialization of his much-loved village.

That same year a more familiar image came about, of gates more distant, more theatrically rusticated. In his 1915 film *The Tramp*, Charlie Chaplin's affecting itinerant splayed his feet for the first time and, with a kind of *plié*, waddled silently, pathetically into a sunset, wearing his wounded dejection as a trophy. His gate was a place not of introduction, but of departure. His unrequited ardency thwarted, the comportment of his leaving could only be expected. The financial futures of the Essanay and Mutual studios were going to come to depend on it.

From the outset, *The Tramp* put in place the recurring details of Chaplin's best-known character. He appears in the film harried by modernity. Beset by malicious or at least uncaring automobiles he immediately plunges into a quandary. The farmer's daughter (played by Edna Purviance) is being pursued by robbers. He saves her, comically, violently, but reveals himself to be as tempted as the thieves he has just repulsed to help himself to the cash she is naively carrying around. The plot unfolds in what would become a familiar way. He escorts her back to her home. Her grateful father offers him temporary employment, at which he is hilariously and cruelly inept. The burglars return and again, exercising his street-wise bravery, he sees them off. We are given other details of his character—his pathological and continual tiny larcenies, for instance—and begin to wonder if these pilferings are intended as remarks on the ethical unreliability of outsiders. We see him disappointed in mistaking Edna's endearing rural innocence for love. And, in the end we see him head off from her garden gate, saddened but also rather relieved perhaps to escape the solid and reliably self-interested society of farmers.

In fact, Chaplin's tramp sounds very similar to the way Fletcher Moss cast his interlocutor at his own garden gate. One thing apparent about Chaplin's character is that although poor and unfortunate, although bedeviled by automobiles, he is as keen as Moss's modern and entrepreneurial itinerant to get back to where he belongs—the road, and the known, anonymous pleasures of traveling for one's existence: traveling as one's existence. What we have in Moss and the figure of the farmer's daughter is the opposite. They represent paired images of the hopes for a settled and decentralized modern culture. The architectural equivalents for those hopes are seen in the approximations of the farmstead that Chaplin designed in order to exhibit his tramp's irreparable lack of fit with an ordered, bucolic life. They are seen too in the architecture dreamed by Moss. As he was writing, the reputations of Barry Parker and Raymond Unwin were becoming both more assured and more attached to a language of domestic and local municipal architecture that was derived from established arts and crafts motifs. In their then well-known development of Letchworth Garden City, battered walls, half-timbering, steeply pitched roofs, latticed windows, and the use of largely undecorated wood and stone as primary materials were the mainstays. So too was an architectural manner of disarming frankness and integrity with these elements. This rhetoric of ingenuous candor repeated itself in civic terms as much as it did in terms of the deportment of architectural elevations and interior detailing.

There are many grounds on which we can make further connections. The type of early industrial Californian farmland that Chaplin worked, and the type of ancient farmland economy that Moss saw his Parsonage at the center of, represents one such. The long, liberal discussion of decentralization that appears in the context of the garden city movement via Ebenezer Howard and H. G. Wells, and which appears later in Frank Lloyd Wright's modern urban *gemeinschaft*, is another.

Two especially significant points, however, are to be made given this comparison of early twentieth-century social desires to stay and to go. The first is the way that the garden gate appears as a contested site of civic sociability. The different kinds of local and broader knowledge, the different types of modern sensibility, the different types of local attachment, senses of history, of trust—in short, senses of place that are articulated between Moss and his capable young modern, Edna and her equally able tramp are emblemized by these gates of theirs. More, the gate here can also be asked to stand in for a host of built appurtenances through which the architectural hope of such animating moments of association are formalized. Stoops, pergolas, verandas, summerhouses, porticoes, vestibules, stoa, porches, pavilions, doorways, garden gates, garden fences: these architectural indices of association also form part of an extended grammar that appeared disdained by modern architectural thought. The "nonceremonial" use of such traditional devices was the subject of only the coyest approval from the moments of its exhaustion by Tony Garnier until a renaissance in neo-vernacular types appeared for mainstream practice in the early 1970s. As seen in Garnier's drawings, the aggregation of civil voices to a genial and indistinctly convivial murmuring is both occasioned by and pragmatically justifies the existence of such architectural forms. That sound, in its dreamed state a rounded hubbub, "ghosts in" the lineaments of the ideal politeness of an ideal social architecture, while at the same time supplying a kind of rose tint for the akwardnesses, anxieties, and misunderstandings that underpin the modern encounters advanced by Moss and Chaplin.

We'll soon see why that resurgence in the neo-vernacular and the idea of the accidents of ordinarily unruly sociability coincide in the 1970s with the early emergence of video art. For what is also clear about these examples is that the acoustic constellation of their social space remains only a guess. The conversations are recorded, recollected only as written text—as Moss's memoirs or as the goodbye note from Chaplin. Pristine in that acoustic regard, they lack as yet any kind of patina or accident, any kind of aural dust.

2 Allophonies: Accidental Architecture

For modernist art practices across the board, the formal command of the accidental remark is a privileged thing. The drips in Jackson Pollock's paintings, the effects of preparing a piano for John Cage's compositions, the junk in Eugène Atget's photographs, the blooms and blushes that appear for sculpture: each of these practices of expectation prepare a philosophical ground for accommodating the delinquent other. This aesthetic hospitality establishes grounds upon which details foreign to the world of a practice might be drawn into accordance with it. Given legibility as an other, such renegade details are asked to behave in ways that justify and reinforce the legitimate outlines of the practice while also seeming to lead away from it.

Announced not as objects but as phenomena of a kind of secondary perception, it is not always possible to give names to such accidents. For the recuperative histories of vernacular architectural elements, those gates and fences and porches at which people meet, became the subject in their different ways of arts and crafts discourse, the spatial-narrative aims of the garden city movement, and the social regulation hoped for by the new urbanism movement.[4] As gates and fences, they are not significant in themselves.[5] Only as ciphers, as sites and occasions of social accident, of unpoliceable conversations, greetings, acknowledgments, and any number of mundane or dramatic meetings do they gain capacity to author space. Importantly, the acoustic blush of sociability narrated by these accidents constitutes the credibility of the architecture as an answer to the questions of polity and integration posed by the architecture itself.

For recorded sound this dust of accident appears in very familiar ways. The crackle and pop, the acoustic patina of vinyl recording caused by particulates, appears as a nostalgic figure. This is an acoustic quality that has been exploited by a host of recording artists to suggest the sentiment of time past. At the same time, this crackle might put us at the threshold of a kind of *glitchwerk*—a music composed of errors, technical defects, and other kinds of unwarranted sonic accident. Acoustic dust can also draw attention to a theology of presence, a *parousia*. The romance, the indexical fantasy of presence that one might weave for oneself listening to vinyl recordings of Maria Callas is thwarted by dust. The sound of dust reminds of the fact that this is a sheet of plastic, a serially reproduced, heavily mediated substrate that is the custodian of a technically realized fictional account of a musical event. Maria Callas will never be in the room with you. Yet, knowing this, the promise of presence is brought before a particular kind of listening. This is not a listening to the dust in a way that cherishes it. Nor is it a listening through the dust in a way that exhibits a kind of aural hygiene. Rather, it is a listening that establishes a dialectical, perceptual protocol whereby a subject is created that listens in both ways; producing a questioning at the site of the listening to and the listening through the dust that decorates, outlines, and announces an aural presence.

———————

Two further incidents, more like glimpses, are both temporally and visually connected to the civic questions hinted at by Moss and Chaplin. They lead the metaphor of dust toward a sociable understanding of aural space. On December 18, 1918, about a month after the declaration of the armistice, philosopher-dramatist-theologian Gabriel Marcel made an entry in a diary.[6] In 1974, for whatever reason, Eleanor Antin did not exclude dust—in this case the sound of buffeting wind and of overflying aircraft—from the soundtrack of her film *The Ballerina and the Bum*.

From his diary, it seems that Marcel had, over a few weeks, become preoccupied with a principle of recollection. He had wondered if recollections were able to conserve themselves, with all the difficult ontic questions of self-repair implied by this understanding. If they are not, he asked, is there a specific activity that conserves them? Immediately noting the awkwardness of such a question—recollections are, he said, confined to psychical and not material spheres—he drew himself back to a further distinction between preservation and conservation. He proposed a metaphysical motionlessness for the preserved as something eternal and somehow profoundly meaningless; delightful, but abstract and lacking entirely in duration—something constituted precisely as that upon which time does not *gnaw*. To the conserved however, he lent a continuing contemporaneity. He held this type of lasting event as something "recalled and coloured by a moving present," a present

1.2 Duany Plater-Zyberk and Company,
Seaside, Walton County, Florida, 1984.
The classic example of new urbanist
development. © Steven Brooke
Studios.

that evokes the event and "arrays it before itself."[7] For Marcel this iterative discrimination was politically important as a threshold of encounter. It essentially defined his attempt to figure the grounds of a departure from the dominance of approaches to religious experience that privileged a required, private, and unmediated encounter with the presence of God above any other form of divinity.

Marcel put his preoccupation with recollection and, as it were, the subject of recollections rather more clearly in the later sections of his journal. In a discussion of one aspect of his love for music, Marcel wrote that he could occasionally perceive a distinction between, on one hand, music that he could appreciate instantly because it fit into schemas that he already possessed (*reminiscences*) and, on the other hand, music that he could only appreciate later, though more profoundly, because it obliged his invention of new perceptual or emotional categories. This latter music, he said, insisted on a "regrouping" of himself. With this new music, Marcel wasn't necessarily referring to the new musics that were surrounding him, such as that of the Second Viennese School. Any new music would do, even; perhaps especially, music heard again, differently, and which prompted a repositioning and a reconstitution of a self.

Christian, though at this point not fully converted to his later developed apprehension of Catholicism, Marcel's inner auralities were unavoidably colored by his knowledge of the transcendent inner aural attentions described by Thomas á Kempis in his *Imitation of Christ*, or those diaried by St. Ignatius of Loyola, Julian of Norwich, Catherine of Siena, or any of a number of Christian mystics who knew personal aural revelations of divinity. However, Marcel's divinity was something as much located in the matrixial character of profane communication between uniquely capacitated human agents as it was to do with conversations between people and God. The civic dimension was its default.

Moreover, his notion of conservation was dependent on the incorporation of the event and its perception within a living, communicating person. For Marcel, the person was the medium of an intersubjective act. For this he elaborated a particular and appropriate view of conversational physicality; of an interposed body, in contact with everything else in a universe; a participating body that sustains the possibility of the *thou* of an heroically uncompromising, intimate *now* in communication; a *thou* that resists the distancing subject, the he, she, and they of the plausibility-lending witnesses to the event recounted in the secular conventions of autobiography. In this attribution of complex civic meaning to the bodily promotion of the *thou*, it is significant that Marcel was contributing only to the most urgent streams of renewed theological discussion. These were streams whose key problematic was found in a critical view of the mechanistic, expectation-establishing habits of secular historical methodologies. One thing is perhaps clear. The uniquenesses and acentricities tied up in the *thou* advocated by the Christian modernism of Marcelian intersubjectivity are not at all present in the modes of meeting and departure recognized by Moss and Chaplin.[8] Whereas the Marcelian *thou* rests on a resignation and delight in the fact that agonists in a conversation are yet to find the complexity of each other's haecceity as a result of the poetics of that very conversation, for Moss and Chaplin a knowledge of the other is politely taken for granted.

3 Plié

Eleanor Antin's video *The Ballerina and the Bum* (1972) takes a similarly nuanced autobiographical course. It is acoustically complex, and like some other morally charged aural-spatial experiments (e.g., Carl Dreyer's *Vampyr*, Robert Bresson's *Pickpocket*, Robert Frank's *The Sin of Jesus*), it unpicks a seam of expectation while at the same time evincing its bodily confounding of that expectation. The ostensible elements of the film's diegetic structure are worth detailing. Edited together from several long takes, the film explores a conversation between a pair of agonists, the ballerina, played by Antin herself, and the bum, played by Lennart Bourin. The film opens with a sentimental gesture toward cinema, cinema's history, and the poetics of cinema's evocations of history. Accompanied on the soundtrack by a creaky recording of a ragtime tune, Antin, in tutu, walks out of the far, sunlit distance, along a railway siding toward the camera. She encounters Bourin, sitting in the doorway of a boxcar. Strikingly, and from the outset, the acceptance of the boxcar and its precincts as currently Bourin's demesne is unquestioned. A civic proscenium to be shared with Antin, it only happens to be formally owned by a railway company.

The character of the bum as articulated by Bourin is one of the politically refractory hero figures of American popular culture. Seen in Harry Partch's score and libretto to his autobiographical *US Highball: A Musical Account of Slim's Transcontinental Hobo Trip* (1943) and played by Lee Marvin in Robert Aldritch's film *The Emperor of the North* (1973), this itinerant speaks to the means, values, and circumstances of a cultural alterity produced by calamities within American industrial capitalism. It is worth bearing in mind that, at these sites, the anti-authoritarian connotations of the architecture of the boxcar is furthered as a forum for the politics of the dispossessed. It also represents a platform for the reflective observation of a landscape. The bleached and dusty terrain of Chaplin's soppy tramp, which is in turn evoked by ragtime, sunset, and Antin's plié, is equally

1.3a The ballerina asks directions of the bum as he rests in his improvised porch. Eleanor Antin, still from *The Ballerina and the Bum*, 1972. © Eleanor Antin/Electronic Arts Intermix. Courtesy of the artist.

1.3b Conversation at an improvised porch, against a background of aircraft noise. Eleanor Antin, still from *The Ballerina and the Bum*, 1972. © Eleanor Antin/ Electronic Arts Intermix. Courtesy of the artist.

colored from the purview of the history of art and architecture by the mute, surreal, modern industrial pastoral of painters such as Charles Sheeler and Charles Demuth.

———

Bourin plays his character with an unembarrassable and bearlike mansuetude. Better fed than the rural transients pictured by Walker Evans and Arthur Rothstein, his bum both possesses and is satirically redolent of the qualities of resilience given to them as ideally worthy subjects of American labor—suitable recruits to

Roosevelt's New Deal. Bourin's bum is not anxious. He brings a relaxed, beer-sipping affability and a ruminatively existential philosophy to his occupation of a specific architectural form. Happiest when casually opining to passing strangers on the consistency of things, this is a character who has had ambitions (disappointed), and who now waits for the arrival of his stake—enough to open a small photography shop in New York.

Antin's film is littered with such small, culturally legible ironies. So, architecturally, while indicating the boxcar as a mobile platform, here immobile and momentarily redundant, Bourin also sits in, and occasionally sweeps out, a porch. With one scatological aside, the territorial hint is underlined as he wanders down a banking, philosophically serving here as a ha-ha wall, and pisses against an unruly hedge. These architectural gestures diegetically serve to locate the civil social character of Bourin's first gestures in this place: to offer to let Antin set up temporary home in the boxcar, to share a beer and a meal, and to engage in a conversation— all unbidden.

During their ensuing conversation, at one missable moment, and despite his actorly self, Bourin notes, with a rolling glimpse, the interruption made by an overflying aircraft. This is perhaps one of the most compelling aspects of Antin's film. Continually there are offstage noises. The microphone is incessantly bumped by the wind. Repeatedly, aircraft fly over, trucks roar by. This to the extent that at many points the improvised conversation between the pair, as they explore one another's hopes and expectations of a modern existence, is almost drowned out. Almost: and this, it appears, is a point to be drawn from *The Ballerina and the Bum*. The film contrives to coincide conversation and aircraft drone as aural modernities, and to construe their imbrication as the structure of a threshold of perception. This is a perception that upsets not only the architectural meaning of the boxcar and its environment as a meeting place, but also the time in which this meeting occurs. To establish a kind of timelessness, the dialogue relies on such anachronisms. In-text references imply that the time is a conglomeration of the American thirties, fifties, and seventies. Similarly, the space is both porch and boxcar, both proscenium and parlor, and in a usefully prescient sense, a *charrette*.[9]

Subsequently, Antin has herself remarked that had she been able, she might have excluded these interruptions from the soundtrack. Hosted as present phenomena by the corpus of the film, these sounds were not then perhaps intended by Antin as substance to her conversation. The space evoked here may have *nonauthored* elements to it.[10] Ballerinas, however, as the film states, remain hungry, no matter what. This ambiguity of the status of what is intentionally and actually hosted by the soundtrack describes the allophonic condition of the film, its dust. The musicality of the conversation of the agonists, the way it improvises direction from

1.4 Beans cooked over a stove improvised from a Folger's coffee can. Eleanor Antin, shot on location during the making of *The Ballerina and the Bum*, 1972. © Courtesy of the artist.

1.5 Lennart Bourin is distracted by an over-flying aircraft while attempting to describe the skyline of New York. Eleanor Antin, still from *The Ballerina and the Bum*, 1972. © Eleanor Antin/Electronic Arts Intermix. Courtesy of the artist.

a lead, bringing forward the *thou* of an intersubjecthood, represents just one of the film's aural loci. But, with Bourin's glance at an intervening, interrupting sound, hitherto hygienically contained by his stoic persistence in refusing to be put off his conversational stride, he folds the acoustic delinquency of the *now* of the circumstances of their engagement into the complex heterotopia and heterochronia of the film's intentions.

Bourin's glance has a canonical visual pretext. This is to be found in mannerist paintings of the crucifixion, pietà, or deposition, where John the Evangelist (who, according to St. Augustine, had heard Christ's heartbeat), gestures toward Christ's wounds while broadly addressing the gaze of viewers, suggesting a desirable cathexis with Christ and community, and effecting an extradiegetic triangulation of spaces and times. What is most marked about this type of Johannine time and space, and the unuttered speaking it allows, is the form of the *now* that it connotes. This most intimate and best-attested phenomenon of Christian mysticism suggests, as Evelyn Underhill has argued, a presence that belongs not just to a historical past, but to a past and present, a past and present that are both historical and mystical.[11] Perhaps unintentional, or rather inadvertent, the formally Johannine character of Bourin's visual slip to the aural, here in the way he accepts the extradiegetic space of Antin's film, also contains a Brechtian distanciating gesture toward an audience. The gesture articulates a frustration with an interrupting otherness, but the frustration is something around which to unconditionally regroup. It appears phenomenally to open up a specific field concerned with the sympathetic curiosity aroused by an other. In so doing, this aurality stages the foundations for a pursuit of an as yet unimaged community, as yet uncountenanced by the film. As dust, that aircraft sound may, of course, be fully attributed to Eleanor Antin's editorial error. Or, it may be curatorially integrated into a critical account of the industrial modernity against which the conversation of ballerina and bum, in its singularity, struggles. Otherwise, it is the token of an other narrative reality, one that provokes a curiosity and regrouping in order to maintain the aesthetic tensions between the types of times and places evoked by the film. This making of space comes in the act of listening to and simultaneously listening through that aural dust of the film.

The way Gabriel Marcel evoked an image from the past as a means of dealing positively with the urgencies of the present makes him important as a pretext for a gestural figure of regrouping like Antin's. What gives his pretext further significance is the way he maintained an image of divine mystery within his frameworks of understanding a radically personalist view of sociability. That motif is one of unknowing. Marcel's sociability here is based in the notion of a reconstitution of the encounter between individuals who know themselves as subjects who do not always already know the other, who make no presumption to an understanding of the other, and for whom the origin of society is the reflective and mutual admittance of a lack of knowledge of each other. It is a form of what has since been called *autre*-biography.[12]

The examination of communities that need to remain small enough to sustain the viability of their particular sociability is a continuing thread in Antin's work. She has said that although she herself pursued an education in philosophy from an atheistic perspective, the kinds of (principally Sartrean) existentialism that she encountered were not foregrounded as the underwriting of the conversational encounters of the ballerina. What connects Antin's proposed theory of urban democracy with Marcel's is the identification of something sustainable only in the unique and ungraspable details of an interpersonal encounter. It is exactly the dust of encounters, precisely that for which no grander, integrationist historical narrative may be provided, that describes the locus of their shared civics.

———

The relationship of Antin's early films to feminist approaches to community during the early 1970s has been fully explored and documented, by Antin and others. However, at the time, that issue of the radically small, resistive viability of the community had clear and influential expressions in other fields. One of these was the latter-day distributist economics of E. F. Schumacher.[13] Another, related intellectually, was to be found the shift toward pastoral rather than doctrinal issues promoted by the conversations that led up to the ecumenical pronouncements of the Second Vatican Council after 1962–1963. Largely disregarded by much secular philosophical discourse, Marcel, like Jacques Maritain, Emmanuel Mounier, Dorothy Day, and also Martin Buber, was one of those whose attentions to the importance of *meeting* shaped the terrain against which the Roman church limned its decision to take notice of long-standing (and fearsomely opposed) modernizing forces within its own body, and to engage with the material experience of modernity. For these thinkers, the encounter between individuals was privileged over the institution, the party, or the organization as the frontier at which the church should understand the spiritual ambitions fostered by modern space.

For Marcel, the body, as the vehicle for the cherished now in communication, with all its implications of uniqueness, was the opposite of the romances of self-certain bourgeois individualism. As his preferred trope, he structured the body dually. The body stood first as a conditional motif for the conservable now by nature of its very frailty and changeability. It also stood as a referent to the mystical body of Christ, the very image of recognition of the otherwise physically and socially dispersed modern Christian community.

Intentionally or otherwise, Antin's film takes part in a laification of this image of community, in very particular ways. Much of her portion of the film's dialogue concerns a lay ballerina's body, which has developed over time and through a critically inspired labor. It is a repository for a relict and disregarded physicality whose only effective connotation is that of a resistance articulable about themes of redundancy and fallibility. It may be resistive exactly because of a lack of any *institutional* desire to appropriate its redundancy. The comic drama of the ballerina is that she is trained, dressed, and ready, and is hitching from California across the country to audition for a place in George Balanchine's New York City Ballet. Her melancholy is that this will not happen, and she knows why. Nevertheless, she still wishes to show something physical to the George Balanchine who required a violent *askesis* of his dancers—the plastic surgeries, cruel suffering, and regimented contortions required to meet his choreographic vision. She wished to show Balanchine something Russian that his cold requirement for austere conformity has played an important role in forgetting, ignoring, marginalizing. In this sense, and in the borrowed architectural context of the boxcar, Balanchine's ballet stands as a metaphor for the real and imagined culturally alienating effects of modernist, municipal urban planning.

Taking up Bourin's offer of a place to stay, the ballerina colonizes a corner of the boxcar, and sets up a little nativity—some ballet shoes, some photographs—and she supplies a narrative. She tells of how she taught herself ballet from a book. She explains that she can't be very good at dancing on stage, as she had to practice in front of a narrow upright mirror. She physically demonstrates a kind of professional deformity, floundering, undignified, while clambering down from the wagon after being invited to eat. "Ballerinas are always hungry."

She explains, too, that her role model is Anna Pavlova, and that she follows the choreographic teaching of Enrico Cecchetti. She points out that Pavlova's style aesthetically celebrated a grammar of individuality, that each gesture was finished with a flourish or grace or a romantic dejection or haughtiness, depending on how Pavlova wished in that performance to articulate the role and, importantly, herself in relation to it.. More, she says that the audiences to Pavlova's performances knew this. *Ad libitum* embroideries were a part of a compact of expectation, now passed from the appreciation of what the ballerina's educated body could plausibly say. This, Antin's ballerina asserts, is utterly unlike the narcissism of Balanchine's new techniques and their reinforcing pedagogy, which demand a mechanical uniformity and concert from dancers, insist on the standardized, rigid conformity of steps, movements, and postures, treating any supernumerary gesture as litter, not

1.6 Antin's ballerina clambers inelegantly
down from the boxcar. Eleanor Antin,
still from *The Ballerina and the Bum*,
1972. © Eleanor Antin/Electronic Arts
Intermix. Courtesy of the artist.

dust, specifically disapproving of, in fact violently disallowing any place for a spontaneous remark through a subjugative incorporation of the individual to the clearly defined shape and intention of the directed troupe. With this, the view given to Balanchine's *corps de ballet* takes for itself the possibility of designating and eliminating error, of twisting an audience's will toward saturated concurrence, and to the subjective appreciation of a choreographic hygiene. With Antin's romantic view of Pavlova's model, however, space is reserved for the construal of the mistake, the informality, the accident, dust, as the vital aesthetic success—a lost and found token of the tenuities of intersubjectivity.

At all points in the playing out of the questionable self-narrative of the ballerina, the soundtrack, by turns, re-inforces elements of what she says and then directs away from them. A sophisticated aural language of bumps and clangs, interrupted music, overheard traffic, and her own speech, both convinced and unconvinced by the self she is trying to bring into being, establish the ballerina as a questionable perception. It is exactly at this point of the opening offered by Pavlova's status as a relic hat the possibility of the conversational *thou* engages with the entirety of the accidental and intended components of Antin's film. The gyres of Antin's temporally dislocated interpretation of Pavlova coincide with the unanticipated sound of wind and aircraft to construe a locus of nostalgia, rather than nostalgia itself. As a locus, it refuses any easily categorizable maudlin quality. In supplying a dramaturgy for the interpersonal architecture of Bourin's boxcar, that sound also presents delin-quency as the details of a personality, physicality, and demeanor in their condition of being nonintegratable. It is the way these physical details metaphorize exactly that on which time might *gnaw*, as Marcel put it, that a discrimination is made between the preserved, expected roles produced by an approach to social space and the role of duration in remaking subjects as conservable subjects of the space that is conditionally constructed between self and other.

Both dressed in the raiment of elegy, but neither Antin's ballerina nor Bourin's bum is recuperated or pre-served by the narratives of entrepreneurial enthusiasm or the ethics of New Deal workerism that are supplied by the film. The significance of the grace of Bourin's Johannine glance toward the sound of passing aircraft, and the embrace of his words by that unwarranted sound, lies in the way it is used to wrestle singular resis-tance from these figures of peripherality, the ballerina and the bum. Utterly lacking utility, overhanging the context of the professional, ideological, and sense-making orientation of the ambition of these figures toward the city, that delinquent sound and Bourin's glance represent the threshold of community.

4 Enclave Urgencies

The interests in nostalgia shown by Antin and Marcel as something mediated and precipitated by views on technology present nostalgia as a site offering modes of association. Even if Antin made a decision not to acquire necessary technical expertise in filmmaking techniques, catching a fragment of aural dust as she did;

even if Marcel could sense a voluptuous shiver in the chance of species suicide by military technology; this does not mean that either of them harbored reactionary longings for preindustrial social structures. They were both interested to find out, through a criticism of mass society, how specific, desirable forms of association can appear lost, and how the form of that loss can offer plausible models of social redemption.

Recently, as an ethico-poetic technology of interruption, a notion of *askesis*, of disciplined self-regulation, has been brought to the discussion of the relationship of different Christian religious organizations to the notions of community promoted by new urbanist planning.[14] American Christian churches have come to think about their historical relationships to urban development. This is in terms of making the most productive social and economic use of urban real estate assets and in terms of the ongoing review of the current viabilities of architectural forms developed in historical contexts where the relationships between church and civic endeavor were more overt. The perceived failure of new urbanist settlements to produce anything recognizable as a materially present community has become an important element of a critique. A view of new urbanist townscape, beautifully comprising all the necessary appurtenances of civic and informal shoulder-rubbing, yet devoid of any of the appropriately inclined subjects required to carry through those forms of association, has become a familiar trope.

As it has appeared in this discussion, the notion of *askesis* has been lent a remedial function. This takes the form of a sacrifice of the desire for quiet, uninterrupted isolation in order to produce the kinds of social *necessity* that enable but are not found in the choreographed forms of association vaunted by the promotional and critical literatures disseminated by such leading new urbanist practitioners as Duany Plater-Zyberk and Calthorpe Associates.

A type of *askesis* figured strongly in the themes of responsibility advanced in Oscar Newman's book *Defensible Space*, one of the early texts of new urbanism. Newman included in *Defensible Space* a photograph of a bit of graffiti painted on a door at a covered entrance to Minoru Yamasaki's Pruitt-Igoe housing complex in St. Louis. It says "Trix Hubert Archie." It certainly represents a graphic trace of a meeting and departure. Behind it might be imagined a sociability of giggling furtiveness or shouting aggression, or whatever else. The graffiti might be applauded, or not. What is easily imagined nearby it, also, is a rumbling explosion. The building was famously demolished in 1972. The graffiti represents accidental iconic anchorage for Newman's book and its diagnoses, and it decorates what is taken to be the gravestone of modernist social planning.

The architectural theory advanced by Newman was simple: when lacking a means of surveillance, public spaces sponsor criminality. His solution was to maximize surveillance, or at least the impression of it, and to minimize as far as possible all unobserved public space, bringing it into one or another order of private territorialization. The architectural devices privileged for this apparently benign regime of visual surveillance were exactly the porches, garden gates, and fences that carried with them the nostalgic sentiments of a lost, hopeful, and peaceable American civic life.

But it was in his concern for acoustic surveillance that Newman was most inventive in technically effecting the private territorialization that is at the center of his proposals. Microphones in elevators, permanently broadcasting both to a central security office and to all the corridors in a block, offered for him an effective means of observation, one considerably cheaper than video cameras. Within the broader framework of his acoustic organization of policeable space, Newman's disdaining attitude toward "audio privacy" was uncompromising; he sought to alter directly the aural discrimination of tenants. Although understandably desirable, he wrote, audio privacy may also be "operating as a contributant to undetected crimes, where it provides excessive insulation of tenants from the corridors outside their doors."[15] He advocated the design of an intercom unit that would be left permanently on and, where its lowest level of amplification corresponded to sounds heard when listening through a window, might assist tenants in detecting the earliest stages of criminal behavior. He held that this process of permanent listening could also encourage an aural familiarity with regular comings and goings, footfalls and accents, and thus help "discriminate strange from normal sounds." Arresting

1.7 Minoru Yamasaki's Pruitt Igoe housing project (1952) as it appeared in Oscar Newman's 1972 book *Defensible Space.*

enough as it is, it was further suggested that the principle of the electronic acoustic window be extended to inter-apartment intercoms. That such a practice has been more generally adopted in terms of looking out for the welfare of frail neighbors, as much as in its potential for monitoring the distinctions between noisy unruliness and incipient crime, says much of the faith Newman had in his subjects and the aurally connoisseurial appreciation of civic territory he hoped to inculcate.

In 1972, Newman attempted to render desirable a detailed aesthetic attention to what would otherwise be regarded as mere noise. His arguments folded theses concerning the formation of visual and aural territories into an account of novel policing possibilities, fostering the sense of security as the imperceptible new stake in the design of municipal architecture. In many ways, Newman's prescriptions have been adopted by new urbanist urban theory and its clients; especially its visual semantic aspects. However, from the anecdotal accounts, a keener readiness may be sensed on the part of the inhabitants of new urbanist settlements to legislate against ambient noise offense than to develop a delight in it, to embrace and be embraced by it. In many senses, as has been influentially argued by Alex Krieger, the time has passed to criticize the forms of new urbanist architecture itself. Flourishing, tied to the municipal ambitions of many American cities, it is an economically successful and socially desired form. Nevertheless, it wears its own melancholies. Denied and, it must be said, definitively denying itself the scope to address the city as a whole, it is tied to an unavoidable enclavism.[16] The much-applauded notion of the transect, for instance, which dates as Helen Rosenau has shown to Baroque town-planning proposals, is intended as a framework of visual context, not a sphere of influence. New urbanist planning has shown itself thus far to be unable, or ideologically unwilling, to address the macroscale issues of dormitorial homogeneity that make for the rhythms of quietness of its environments.

Moreover, as has been noted before, it is not possible to complain sensibly about new urbanism's stated aims to be "diverse, compact, pedestrian and celebratory of the public realm." It is not possible to decry hopes for mixed cultural and economic structures, high densities, short walking distances, decrease in car usage, or a distaste for suburban slovenliness, and a more generalized praise of urban living. How these things have either failed to materialize or come about only in limited forms is another matter. It is certain that if any principle of voluntary lifestyle *askesis* is to remedy these shortcomings, it is a principle that must engender a renewed form of geographic sensibility to the regionalist ones currently prevailing in new urbanist thought, one that will countenance the kinds of macroscale adjustments in employment patterns needed to enliven daytimes and late evenings in Seaside, Celebration, and Orenco Station. It is this that Lennart Bourin's rolling, skyward glance toward a delinquent acoustic porosity perhaps best foresees. If the aural sociabilities of new urbanist settlements are, like those of Charlie Chaplin and Fletcher Moss, available only promotionally and nostalgically, as written text, if new urbanism's perceived failures to engender informality about the architectural types it repeats and rearticulates are genuine, if its advocacy of nosy proprietorialism as the means of precipitating a sense of security has eventually to turn to private security firms, intercoms, and electronic gates—then it is only its dust, the extensive and volatile discussion *over* the subject of new urbanism's lessons for living that represent its contribution to community.

2.1 Advertisement for Imhof Radiograms,
     ca. 1936. Image courtesy *The Gramo-
     phone.*

2 Omissibility

But, how many things do we engage in without

being certain?

—Lucien Goldmann[1]

The knowledge gained by the act—rather than the content—of radio listening is a particular kind of knowledge. Here we'll see whether a musical device, the hemiola, can function as a device for interpreting modernist architectural history.

———

The advertisement shown in figure 2.1 is fascinating as a charged depiction of suburban listening. Even though it is clearly staged for commercial purposes, a kind of coy romance seems to be emerge in this image of prim attention to a score and a radiogram. Or perhaps we can see some malevolence: looming, barely suppressed. Or, in fact, we can see anything else we may care to bring to it. This is an image of aural-architectural domesticity that could be said to be baroque—emotionally promiscuous in what it may contain. Produced in 1936 as part of an advertising campaign for Imhof radiograms, it showed up variously in the popular press. It appears in the more specialist publications of the period, like Novello's *Musical Times*, Compton MacKenzie's the *Gramophone* and Jack Payne's *Popular Music and Dancing Weekly*. It was seen in the *Listener* and the *Radio Times*, produced by the BBC, and in the dailies. It circulated widely, and because of what it *may* represent as a fledgling or already established relationship, one whose details may not have borne too public a scrutiny, it is an image possessed of a kind of revealing ambiguity. Is our man here a potential assailant? Is he listening, disappointed in love, about to ask for the car keys?

The subject of uncertainty has become an illuminating mechanism for those interested in history, whether it be the history of film, music, or architecture. Suspicions, dubious irresolutions, indecidability, and the utop-ics of intangibility, imperceptibility; all have helped shape the recent conventions of historical accounting for cultural forms. What follows here concerns the articulation of the historical variety, complexity, and uncer-tainty of modern, suburban temporality in Britain in the 1930s. In part, it is a historiographical inquiry, one that broaches the likeliness of particular audiences for the writing of "modernist" cultural history, as well as the role of those definably suburban audiences in organizing varieties of significance for such histories. As an inquiry, it concerns the technical abilities of these mooted audiences to think about and between architectural and musical forms, as well as prevailing kinds of historical narrative convention, and acceptable directions of architectural discourse.

1 Weak on the Weak Beat

Nikolaus Pevsner's account of the ontogenesis of the preferred compositional forms of architectural moder-nity, *The Pioneers of the Modern Movement* (1936), is probably his best-known work. It could be readily read against another contemporary and equally pedagogically influential work: Arnold Schoenberg's *Fundamentals of Musical Composition*.[2] These books, their subjects, and their products may seem rather at odds with each other. However, they do seem to share attitudes toward the historical emergence of modernist aesthetic and cultural forms, as well as toward their critical interpretation and furtherance. Among concerns with vernac-ularity and other issues, they both broach the cultural historical significance of *omissibility* as a founding modernist aesthetic principle. They both also present arresting images of time and of temporal development. Moreover, when Schoenberg, for example, came to discuss the traditions of applied, ornamental musical meter, he castigated it by likening it metaphorically to the fraudulent and imitatively incoherent deceits of the master builder's stucco work. One can almost hear Pevsner applaud at the idea. These commonalities are only illustrative, here, of the kinds of shared issues that lay between musical and architectural discourses in the 1930s—the heroic moment of both modernist cultural emergence and the formative accounts of its understanding.

The contemporary audience to Nikolaus Pevsner's arguments may not have had Schoenberg in mind when reading *The Pioneers of the Modern Movement*, or any of Pevsner's other and various articles that appeared in the *Listener* or in the *Radio Times* and elsewhere at that time. Importantly, too, they may or may not have had in mind the rude and reductive alien that was later coined as the reputation of his intellectual probity by other architectural commentators, such as John Betjeman. Betjeman, the poet-proponent of an affected, choleric, studiously disenfranchised, conservative, and offendably English masculinity, presented his own amateur elegism as the preferred subjective platform for the interpretation of architecture. He had a poisonous affection for the idea of a professionally unsentimental and proscriptive "Herr-Professor-Doktor," a foreign and scientific anybody who "wrote everything down for us, sometimes throwing in some hurried pontificating too, so we need never bother to think or see again."[3] Pevsner's larger audience, on the other hand, many of whom incidentally may have been in no great hurry either to see themselves too clearly written into what Betjeman saw as the bland convulsiveness of modernity, could have known and preferred Pevsner's poetic way with the modern. If they held him in their thoughts at all, they may have known and preferred exactly the kind of inexplicit suggestiveness that characterizes his writing, and which Betjeman helped to obscure. Such audiences may have known and preferred his legibly differentiated, overlain and dissonantly allegorical, architectural temporalities. More importantly, while Schoenberg may not have figured for them, though the possibility remains that he could have, Pevsner's audiences may still have known and preferred him and his interpretative devices, musically.[4] The picture of a suburban listenership as a culturally and historically specific, mediating constituency, a kind of inconstant constituency of those who were "most dependent on and welcoming of radio," is necessary here—one might almost say *functional*.[5]

———————

While Schoenberg had things to say about the place of the sentence in the expectation-confounding, unsymmetrical, prosodic structure of modern music and the technology of its criticism, it should be noted too that the poet Robert Frost also once presented an abstractly aural basis for critical evaluation. Frost wrote that the most important thing that he knew was that good readers and good writers read and write with an ear. Each sentence, he said, has a sound, a sentence-sound. Words, as other sounds, may be strung along it. It is an intriguing idea, a kind of suggested though also apparent sound, and one simultaneously comprising but not constituted by other perhaps contrapuntal, and in any case, different sounds: words.

Frost didn't seem to imply an attempt to foist some personal mode of critical mysticism by proposing this notion of the carriage of poetry. It was presented simply, more in the way of a newly regardable, material fact of literature. He said that there were different kinds of sentence-sounds: some common, some mundane, some banal. Some, however, could articulate spontaneity, uniqueness, or some recognizably genuine authenticity, one uncompromised by lazily formal academicism. Certain sentence-sounds, then, may convince the hearer of the revelation of something hitherto inarticulable by writing's competence. The more striking the sentence-sound, the more estimable the writer, Frost proposed. It is possible that he took this strikingness to be some kind of autographic guarantee, something caught up in a fetishism of personal and formally unique expression. To think so would be a pity. His conceit, with some ill will, can also be understood as formalist fogging, but that need not be necessary, either. In fact, Frost was quite clear in his opinion that no writer can legitimately lay claim to the invention of these sentence-sounds. It should be a plebiscitary and representative sound, then. In this, as with other apparently modernist critics at the time, he indicated that he sought the anonymous and current vernacular as the locus of good design.

Most arresting in what Frost had to say about the audibility of sentence-sounds is that it is not a matter of rude, acoustically perceptible facts. Sentence-sounds, as he understood them, evaporate literally at the tip of the tongue. They are *heard* better than they are *spoken*. And what he referred to as "oral practice" simply pushes this sound away. This figuration raises a noetic issue concerning the nature of abstract hearing that is bound to reading. Frost's sentence-sound is conceptual. Though prompted by words, it neither inhabits the same interpretative world as vulgar semantics, nor may it be isolated for general inspection by physical acoustics. It may be dreamed.

The idea of a sound that is physically intangible—sound, if you like, conceived figurally from outset to end—occurs frequently in suggestive commentaries on the way that poetry might make meaning beyond the simple assemblage of words. Often, as is the case with T. S. Eliot, the figured sound is connected to some consideration of the way that a history may be seen to emerge musically—that is to say, both in a text, in the posed formal internal history of a discrete work, and in a culture, in a desire to present some observable and imaginable tradition or development. A poetic concern with rhythm could be useful, then, in developing different perceptions of contiguity and inconsequence in argued historical meaning. Attention to figured sound could promote the recognition of different types of interpretative agency that could amount to a syncopation of history.

A conception of the acoustic figures of (architectural) history derived from poetry and its criticism is not exactly elite; it is only the thirstily self-promoting Eliot we are talking about, after all. It is, however, certainly an aetiolated one. A perhaps more likely set of resources that suburban audiences reached for, in attempting to grasp the nature of the aural as well as the architectural cultural landscape that they inhabited, was much less formal. And, it is to these more informally articulated pedagogies, such those already mentioned, that we shall more properly turn. Any of a number of dance music magazines and any of a number of introductory guides to the appreciation of the classical orchestral and choral repertoire were more familiar to these suburban audiences. This audience gleaned what perceptions it may have had of parsable metrical dissonance from this more informal literature. Precisely this more popular literature, as we'll see, provided access to formal ideas of the apprehension of meter and rhythm in relation to the greater argumentative design of a text as a whole. Notions of discrete and differentiated rhythms overlaying one another and causing such effects as structural ambiguity and imbroglio, or metrical feints, were available in this literature, too. These works also included a recognized possibility of implied contrapuntal argument, and the intriguing theme of revenant or accompanying voices. It is a richly imaginative literature.

In 1936, H. H. Wintersgill published a beautiful short study whose observations are in some ways quite valid today. Entitled "Handel's Two-Length Bar," its subject concerned the most characteristic of Handel's style, specifically his use of the hemiola and the distribution of three beats over two measures in triple meter.[6] In the context of the leading characters here, it is interesting to note that he also made much both of the fact of Handel's presence as a German in England, and of the apparent Englishness of the hemiola. Wintersgill's essay was published at a time when many writers were attempting to similarly detail profound historical mutualities in English and German culture. This was a minor literary genre to which Nikolaus Pevsner himself contributed.

I'll be clearer about the meaning of the hemiola, and, initially at least, try to treat it in ways distinct from those that have caught it up in Wintersgill's early national cultural ideas. Musicology is a discipline possessed of a rich diversity of abstract figurations of history. For some species of musicology, the history of music might mean a particularly observed account of a particular tradition: from Bach, or Beethoven, or Korngold, or whomever. A detailed history of a specific piece of music, however, might just as easily be something aiming to record the social, cultural, and technical circumstances of its production. To speak of a piece's history might be to refer to the curriculum vitae of its reworkings, performances, recordings, editions, interpretations, and criticisms. Each of these versions of anecdotal history is, however, largely supplemental to the requirements and ambitions of analytical musicology.

A piece of music is also understood to have an *internal* history or, rather, several potential internal histories. This history concerns the teleological statement and development of its subjects, strophes, motives, and other components: the *telemes* of the piece that realize its *teloi*. It is this view of musical history that provides the cherished object of formal, analytical musicology. This history is abstract and also, as indicated, considered largely without significant regard to the possible, extramusical social interpretations that may be made from it. It supposes a synthetic, specifically musical temporality, one that is composed dually of the durative experience of the unfolding music *and* in an instantaneous view of its entirety. Quite score-bound, even sentimental senses of historical emotional witness are rejected by it. The technology of conceptual language that has been used to articulate the astonishing variety and complexities of these internal histories, while for some a relic, is nevertheless remarkably accomplished.

There are moments in the internal history of a piece of music that may be taken to articulate and define the nature of that specific musical-historical process. The hemiola can be construed as one of these. As a specific metrical device, a hemiola is effected by establishing a specific meter and unexpectedly destressing a particular accent. This precipitates a kind of crisis in the perception of continuing or emergent rhythm. The effect is often regarded as a sudden slipperiness or cloudiness or, paradoxically, an uncontrollable tumbling, forward motion, together with a sense of stasis. Terms such as *imbroglio, double-entendre, anacrusis, ambiguity, slyness, surreptitiousness, deceptiveness, incongruence, ungraspability, confusion, instability,* and *enigma* have all been used to describe its manner.

The literature on the hemiola is struck through with accounts of its *irresolving* effects. It has been described as a "dislocation of the metrical surface" of the music.[7] This dislocation is something that might be thought itself to figure the idea of the surface of the music in the first instance, and to figure it as a result of articulating, in the same disclosing gesture, the eddying confluence of other, perhaps larger currents moving beneath it. It is a rhetorical device, and one that is technically dependent, in part, on the musical-textual (or other) entrainment of the listener toward a particular metrical expectation, and the confounding of that expectation. As important as the fact that it is regarded as a significant element of the English Baroque is the fact that the hemiola will evince on a score, but not necessarily acoustically.

The point about the observable manifestation of a hemiola is located precisely in its cultural circumstances. Hemiolas are sometimes awkward to achieve technically. Moreover, they don't always present themselves entirely frankly on a score sheet, as the recent literature has noted. Hemiolas may be implied; they may overlap; they may result as an effect of large, architectural, extensively hypermetrical concerns or as a local detail of applied ornament. The interpretive and perceptual skill of the conductor is as important as the proficiency of the players in the successful articulation of the hemiola. Technologically they are often difficult to apprehend, too. Poor concert hall acoustics can lose them. And, since the pertinent rhythmic components may be carried in more than just one part of the overall dynamic range of the sound picture, it follows that poor recording, broadcasting, or reproduction technology may easily lose a crucial detail and, with it, the fugitive hemiolic effect altogether.

The issue of expectation here is of particular concern. The structure of a hemiola is such that in order for the listener to be usefully and pleasurably upset by the lifting of an expected beat, he or she must be sympathetically entrained. This entrainment can be grasped only through attention to an overt (or syncopated) establishment of a principal meter.

It should be noted that much musical meaning may be generated by the labor involved in simply ascertaining a principal meter in any given piece of music. As a corollary, inattention or, importantly, the differing construal of a principal meter can mean that hemiolic effects will be either missed, or, rather more interestingly, even found where they have not been intentionally structured. This is an issue of the physiology of rhythmic cognition, but the point has also been made that such perceptions are in no way universal and are in fact thoroughly encultured.[8] It may be perfectly possible to hear a hemiolic effect yet simply fail to lend it any significant inflection because one is culturally unaware of the likelihood of such an acoustic event representing a legitimate compositional feature in the narrative or philosophical organization of Western music.

One further point is that entrainment may be effected in relatively localized metrical details, but it also implicitly involves the whole of the work—one might say much of the canon of Western orchestral and choral music, as it is these that work toward a listener's propensity to attempt to establish an identifiable principal meter.

It seems, then, that within the structure and dispersal of the hemiola, we have a suggestion of an always ambiguous and, even when discernible, enigmatic moment. This is comprised in the understanding of the confluence of differentiated historical, self-consistent narrative meters as a completely contingent, but nevertheless legitimate, articulation of contradictions in the development of the internally historical, thematic materials of a piece of music. The hemiolic event may be recognizable, or it may not. It may be recognized as one thing, or it may be recognized as another. It is entirely uncertain. As a historiographical morphology, however, the hemiola is compellingly suggestive and might tell us something of the sentence-sound of Nikolaus Pevsner.

2 Pevsner's Pedals

It is worth remarking on the degree of rhetorical certainty that surrounds apprehensions of Pevsner's modernist intentions. The main cognitive part of his influential model of architectural history is founded on the speciations he introduced. That influence comes with a burden, however, and Pevsner has been regarded as aggressive, singularizing, and proselyte, unafraid to distort a historical picture through omissions and weighted interpretations in order to promote a model of the "correct" forms and ambitions of modern architecture. Such malicious questioning of Pevsnerian scruples has been a popular British cultural pursuit for something more than sixty years. What could be its basis?

Much in the same way that Schoenberg is thought to have argued for the natural inevitability of the kind of musical methodology that he proposed, so Pevsner is said to have argued for a particular architecture to be both the inexorable necessity and the *summum bonum* of the modern historical epoch. Of many arguments that he proposed, concerning many architectures, Pevsner's most sustained voice is perhaps the one that places the architecture of Walter Gropius exemplarily at the beginning of the modern tradition. In *The Pioneers of the Modern Movement*, he has Gropius's 1911 factory building at Anfeld am der Leine as the archetype of the true and appropriate international architecture of the modern period. This kind of presentation of an instant of German culture as the corporealization of an imagined international zeitgeist is familiarly Hegelian, and has something of an integrationist and *étatist* aspect. In this historicist manner, Pevsner provided a lineage for Gropius's architecture, which many have seen as a selective and brutal reduction of a diversity of instances of architecture and design to a single functionalist *nisus*.

His perception of the frank expression of materials, structural features and the single-mindedly resolute organization of space in Gropius's factory is understood to commit all those preceding and diverging examples of architectural detail, texture, and structure that he evidentially conjures, to the status of mere photographic *telemes*, whose sole historical existence, whose *entelechy* is only to occupy necessary places in the description of the emergence of Gropius's buildings. But the rhemic subtleties of Pevsner's text are rather more resistant to summary than this.

In some sense, it isn't surprising at all that Pevsner should be thought of as pursuing a monothematic historical agenda. At times, he drew himself as a proponent of a particular and increasingly widely read variety of historical semiotics, Panofskyan iconology.[9] The method of this tripartite analytical approach is described thus: first, the identification of different kinds of signifier in any given cultural artifact from any supposed historical epoch (a drawing by Michelangelo, for instance). Second, the *iconographical* ascription of those signs to one or other stories or allegories available at the time, as a preliminary interpretative ground. Third, the *iconological* ascription of the sign, through its narrative referent, to a personification of the worldview of an epoch, "qualified by one personality and condensed into one work."[10] It is this that presents the true, principal meter against which all else gains its character, its sense, or its anomalous nonsense.

For Panofsky's system of historical personalizations, Michelangelo was the person who characterized the Baroque, and characterized it as the emergence into the modern period. It is interesting to note that where Pevsner made his clearest public alliance with that image of the moment of presumed singularization of modern historical perspective was in a 1928 article on the Baroque. This was at just the time when he was involved in the research that led to his 1930 lectures at Göttingen University on modern design, which contained the arguments of *The Pioneers* and which supplied Gropius as the character of modernity.[11] Approving of Panofsky's method, he countenanced the idea of the characterization of the Baroque and the characterization of the modern simultaneously.

When it comes to the resolution of *The Pioneers*, things are not as clear as they may seem from its ico-nologizing context. Even the most schematic *précis* will recognize that Pevsner's text is structured by at least three reprising strophes. There is a Corbusian notion of engineering design conceived of as problem solving and founded in an unsullied ignorance of cosmetics. There is one sponsored by the use of vernacular, English domestic idioms in a remedial relation to German industrial design, *pace* Walter Rathenau and Hermann Muthesius. The third is a figuration of the *enablement* of modern innovation through descriptions of break-throughs in materials science, advances in concrete technology, and the like. These strophic burdens may be thought of as the different conceptual meters of the text. Representing discrete and differentiated historical speeds and evennesses of emergence, they are used to introduce the wide variety of Pevsner's historical exam-ples. They reprise and intersect with each other, and, at those points of intersection, Pevsner is able to make specific, significant remarks.

Which of these many remarks actually qualify as significant depends on the decisions made by the reader of the text, part cued and part extraneous, as to the structural *vraisemblance* of the narrative. This is because, aside from being a functionalist tract, *The Pioneers* is easily read as an anti-Fascist, pro-Weimar allegory. Sus-picions of socialism in Pevsner's thought aren't groundless, and the subtitle of the book is, after all, "William Morris to Walter Gropius."

This might all be a little more personal, however. Pevsner left Germany in 1934 and secured a research position in Birmingham, where he drew together the arguments for *The Pioneers*. It is also, incidentally, here where fellow Baroque historian Ellis Waterhouse was working at the time; and it should be noted that Pevsner was later, as editor, to commission Waterhouse, with whom he disagreed publicly at times, to write *Painting in Britain 1530–1790* for the monumental Pelican History of Art series.[12] Pevsner's family didn't manage to get out of Germany until 1937. It may be that, sensitive to the situation of his family, Pevsner felt the need to allegorize any discontent with Nazism. It is certain that no modern credence is lent *The Pioneers* to classical architectural modes, the favored civic vernacular of National Socialism. For Pevsner, iconologically, it was the culture of the Weimar Republic that produced Gropius and the Bauhaus, and not that of the Third Reich.

———————

Pevsner is thought to have changed his religious affiliations at least once during his life, converting from Judaism to Lutheranism at the behest of his wife. This could mean anything: a profound fideistic change, or a domestic comfort, or prudence in the face of terrific persecution. It could mean all of these things, some, or none. Nevertheless, it is a consideration that throws *The Pioneers* into a further light. A theological dimension, and the possibility of Pevsner's relative diffidence toward the spiritual over the intellectual aspects of it, bears influentially on the way that the identifiable strophes of the text structure the *teloi* of the argument. If it is viewed solely as a resolution of our trivia of differentiated historical meters, in the singular image of Gropius's factory, then the dénouement of *The Pioneers* is a curious affair.

The passage, for instance, where Pevsner finally heralds Gropius's work, is couched in the terms of pre-modern, Catholic traditions of architecture. He offers thirteenth-century Gothic precedent: "Never since the Sainte-Chapelle and the choir at Beauvais had the human art of building been so triumphant over matter."[13] The potential for such theological suggestions is great. The term *functionalism* has a notable presence in developments of Gothic Revival aesthetics in nineteenth-century Europe, anyway, and Pevsner could also have meant to conjure a modern and secularized form of guild socialism. As possibilities, these are more or less orthodoxies. He may also have wanted to suggest a mystical, alogical, or divine character for Gropius's work. As one such paradoxical remark among many, his comment is odd; the more energetically Pevsner goes on to disallow a spiritual dimension to his conclusions, the stronger that suggestion becomes fixed.

It is worth noting that, at various points, especially in his studies of the Baroque, Pevsner was interested in the manner of the public articulation of inner spiritual life; but despite this, there is little point in psycho-biographically insisting that any of this is evidence of his feeling his way toward a critical mysticism and a Catholic God. Nevertheless, the possibility remains that within the framework of a perfectly reasonable and respectable, secular academic interest in Catholicism as the defining nexus of philosophies that he and others saw as founding the European cultural forms that he chose to study, Pevsner may have developed a set of allusions that allowed him to articulate a kind of sympathy with dissenting cultures. His primary academic interest was, as suggested, in the study of the Baroque, which in Britain anyway was then considered a dissenting culture, and he was more specifically interested in the Catholic cultures of southern Germany, which at the time of his writing appeared to be both coming under the threat of the Third Reich and making their accommodations with it. This rather unfamiliar way of understanding his cultural commentary on modern architecture is defensible. However, its defensibility depends on a view regarding whether or not, in *The Pio-*

neers, a text that is generally considered of exemplary, severe, and assertive clarity, Pevsner desired somehow to preserve a sense of mysterious irresolution. Such a view would depend on a form of readerly reentrainment concerning his supposed polemic interests. It would suggest a kind of overly sophisticated, shall we say Baroque, allegoresis. But the implication of a kind of intellectual sympathy toward Catholicism on Pevsner's part might also make a kind of social and cultural sense, especially to the audiences of more widespread varieties of popular culture in Britain.

3 Pseudonymy

So there is still further context, context that more fully reintroduces the interpretative abilities of an imagined suburban audience to Pevsner's writing on architectural modernism. Elsewhere, I have attempted to locate *The Pioneers* in a different popular context, that of the most pronounced feature of landscape of the period: crime fiction, the Golden Age of the English murder.[14] At that time, normative systems for crime narratives were legislated by such professional mutual interest organizations as the Detection Club of London. These rules established the grounds for a contract that sought to deter writers from making fools of their readers. They legislated over the revelation of evidence, disallowed certain late-in-the-day appearances by supernaturally gifted characters, and demanded a clear resolution of the mystery in the closing sections. They treated the examination of murder syllogistically, like a conundrum. The rigorous, near-Thomist logic presented an idea of a text that could be resolved legitimately in only one way. Practice, however, confounded theory. In fact, many of the writers in this circle consciously went out of their way to break those rules. G. K. Chesterton's Father Brown mysteries are the clearest examples of texts that provide resolutions tenable only if some point of Catholic faith is accepted.

I have suggested that the nature of this modern literary contract between readers and writers was also a suburban one. Crime fiction was one of the literary staples for new, suburban, commuting audiences. The idea of a suburban population suspicious of transient visitors to their new urban environments is an attractive one, too. So, the crime novel, even in its strictest, "locked room" form, meant more than just a teaser, entertainingly resolved, or unresolved. An audience to Pevsner's *The Pioneers* may have already been highly skilled in reading for imbroglio and irresolution in texts composed teleologically of unfolding evidence. They may have been suspicious in their interpretations and tended toward holding probationary attitudes toward any too glib a concluding gesture. It would be possibly too deterministic to insist that these people had mortgages and, therefore, complex, historically subjective investments in teleological closure. But the crime novel may have provided a mechanism whereby, in their diversity, members of a suburban readership could have read themselves into and out of the tangible instabilities of an international, economic, and cultural modernity that periodically appeared to threaten the tenability of their newly and only partially secured roles as members of a nascent, property-holding, popular democracy.

All this suggests something that brings into relation aspects of relict Catholic culture, articulations of modernity, different social habits and rituals, different creative-cognitive enigmas in popular literary forms, and an understanding of a suburban audience as something that was entirely coy about itself and its romantic place in modern culture. While this supplies an attitude of reading that may have been taken toward *The Pioneers*, you will notice that a biography of Nikolaus Pevsner's authorial intentions, no matter how compellingly appealing, could barely figure at all in such a reading, let alone decisively.

———

It may seem unlikely, but despite his hard urban reputation, Pevsner was rather interested in private, suburban architectures. The idea of a suburban location for modern domestic architecture should be no surprise. It is often noted that, where it did exist as more than paper proposals, much of the modern architecture of the 1920s and 1930s existed mainly in isolation on the private estates of individual patrons and in the suburbs of Europe's major cities. Nevertheless, aside from these episodic occurrences of modernist adventure, Pevsner took time and care over other architectural forms of suburban domesticity.

He eventually used his editorship of the *Architectural Review*, the crusading journal of architectural modernism in Britain in the 1930s, to exercise his surreptitious interest. The *Review* published a series of articles, entitled "Treasure Hunts," in which the historian Peter F. R. Donner maintained a celebratory account of his ongoing researches into suburban building types. He itemized ideal porches, chimneys, mullions, and gables, even the special flora and street furniture of suburbia. He spotted Caroline and Georgian features. He watched for Italianate and other national influences, and he assembled them loosely under the headline of an attempt to divine a Coburg-Windsor idiom, one appropriate to the age. This turned out to be a fragmentedly eclectic, contemporary national style, a consensual language of romantic individualism.

These suburban inquiries were noted by such an internationally eminent promoter of the formal appreciation of European modernism as Henry-Russell Hitchcock, who knowingly recommended the example of "Mr. Donner's little exercises in the London suburbs," especially to those interested in the development of national and regional architectural forms.[15] Hitchcock's archness, and his deviation from what has come to be expected of apologists for the modern European architectures of the 1920s and 1930s, may be explainable by the fact that Peter F. R. Donner was Pevsner's own pseudonym. He seems to have borrowed it from the German Baroque civic sculptor, Georg Donner, whom, at that time, he thought simply the best visual artist that there had ever been. It may be possible to suggest that, willfully, and against the general impression of his reputation as a modernist historian, Pevsner made some kind of shy address to the subjects of a usually vilified concept of a suburban architecture, and did so through autobiographic identifications with the idea of the Baroque.

This bears directly on *The Pioneers*. As Donner, Pevsner made a modern plea for domestic historical eclecticism. Again, given the kinds of imbroglios we can now find in Pevsner's works, this could mean the confluence of many things. One of these things is a faith in the approach to culture argued for by aesthetician of the picturesque and favored theorist of Pevsner's, Richard Payne Knight. Over the years up to 1942, the connoisseurial, associative principles of the picturesque, which Pevsner played a great part in elaborating and bringing to a wider audience, came to intrigue him more and more. Viewed from this perspective, *The Pioneers* seems less interested in Gropius's functionalism and more concerned with shaping the image of modern architecture around the domestic, suburban historical eclecticisms of the architect Richard Norman Shaw, someone who occupies a rather privileged place in the text of the book.

This seems to indicate a preferred mode of architectural interpretation on Pevsner's part, one based on a knowing and informed game of historical association and allegorization, played with the textures, details, and plausible symbolisms of architecture. Here the architectural object is never thought to mean just one thing. Its meaning is never static; historical and intellectual sophistication is the rule. This is something that may quite radically complicate the meaning of Pevsner's modernist *teloi*, but it still doesn't justify any insistence that Baroque devices like hemiolas organize them. For that to make any sense, there needs to be some further kind of agency.

4 Klang (The Sound of Music)

Until now, I've only suggested the *noetic* significance of phantastic sonorities. At some point, sentence-sounds and hemiolas must come into morphologizing relation with the idea of a material acoustic in the popular, associative social and cultural constellation of modern suburban space.

In the introduction, I noted that Rudolf Arnheim, writing about the radio in 1936, claimed that rather than harmony being fostered by public music, we should perhaps more readily expect conflict. At a time when loudspeakers in suburban back gardens, letters of complaint about "radio smoke," and moralizing arguments about the iniquitous effects of broadcast jazz music on adolescent girls were novel moral furniture of cultural modernity, the idea of a highly figured kind of socially constellative, national acoustic geography is bound to be a complicated one. The kinds of tussling that showed in the letters to the popular press, where complaints were less about radio sound per se and more about a neighbor's loud enjoyment of broadcasts of recitals of Bach's music interfering with another's enjoyment of the broadcast of Henry Hall concerts, seem to bear out Arnheim's observation. In this warring context, the sudden availability of large amounts of air-time dedicated to the classical repertoire caused its own problems, especially in 1936. In 1936, the departure of Edward Clark from the BBC marked, though most likely did not account for, a shift in the BBC's policy toward the broad-casting of classical music. The schedules at that time show a move away from the innovative and enlightened programming of contemporary, often overtly modern music, overseen by Clark, toward more recognizably canonical music: Bach, Beethoven, Brahms, and especially music deriving from the elegiac, somewhat nation-alist landscaping of Moeran, Vaughan Williams, Bax, and even Rutland Boughton.

Despite the BBC's lessened interest in the productions of international contemporary musicians, one international contemporary figure, Jean Sibelius, nevertheless became rather prominent. Suggested as a north-ern European precursor to the nationalist romance of Ralph Vaughan Williams, Sibelius's music was hugely popular in Britain. Various attempts have been made by commentators to explain this popularity. Theodor Adorno has recounted a likeable discussion with Ernest Walker. Adorno said that he couldn't understand why a musician who "combined meaningless and trivial elements and alogical and profoundly unintelligible ones" and who "mistook esthetic formlessness for the voice of nature" should be admired in Britain.[16] Walker replied that it was precisely these values, reviled in German musical-philosophical culture, that were cherished by the British.

The discussion around Sibelius was shaped by a number of aspects, and not just those themes that bore on his position as an international justification for a renewed attention to parochial cultural forms. His reputation developed in the context of already established enthusiasms for Scandinavian architecture and design, for instance, which seemed to connote, for some, a kind of clear, reasonable, and wholesomely healthy modernity. Moreover, the idea of Scandinavia was important in arguments about Britain's place in a northern European culture, rather than one that sprang from Mediterranean origins. These arguments were widespread and pervasive, pursued in both the popular and the academic press by Herbert Read and others.

However, what is most striking in this contextualizing of Sibelius is the level of sophistication that the popular, pedagogical discussion of choral and orchestral music reached, especially with regard to the practice and rituals of radio listening. The images of advertising for radios, or indeed for furniture, gramophones, and other domestic durables, suggest that radio listening was something to be taken seriously. One recurring image is of concerned, almost fixatedly attentive listening. While jolly, avuncular images of light radio are plentiful enough, an intense earnestness also undeniably figures in the iconography of radio listening. Whether alone, as couples, or in an extended family, the radio audience was figured in terms of a committed eagerness to understand what was going on. An augmentation of this image of domestic listening came with the figuration of the sound engineer. Crisp young men, suited, in headphones, and wearing efficient haircuts, these were the modern acoustic artists: virile and earnest while at the same time somehow clerkishly responsible and dependably administrative. Scrutinizing the broadcast aural picture, they sat with sensitive fingers hovering over dials that monitored and controlled exactly which portion of the technically available sound would go out to the nation.

This notion of the administrative mediation of the technically available sound was no mystery, nor was it isolated as the privileged domain of technophiles. The *Radio Times* and the *Listener*, among others, regularly published articles by engineers such as P. P. Eckersley, and made it clear that since the recording equipment used in live broadcasts could capture only a fraction of the dynamic range of the orchestra, someone had to be responsible for the placement of that acoustic window most appropriately during the playing of a piece that might, in the course of its performance, exercise the whole of the orchestra's range. These short articles on the function of balance, tone, and color controls rarely failed to evoke this partiality. Both visually and in terms of imparted technical information, we can see that a further compact was envisaged between, on the one hand,

2.2   The dynamism of BBC Engineer.
       From Rudolf Arnheim, *Radio* (London:
       Faber & Faber, 1936). Image courtesy
       H. M. Stationery Office and the GPO
       Film Unit.

the sound engineer as a young, modern, suburban individual with a kind of moral responsibility for the cultural welfare of the nation, and the equally committed, increasingly suburban listenership, on the other.

It is worth recalling again here Adorno's diagnosis of the supposed effects of such technological deficiencies on listening skills. For him, the partial nature of the acoustic image of live music made available by broadcast radio directly resulted in an eradication of all the aspects of timbre and tone that in turn disintegrated the sonic and philosophical corporacy of a piece of music, reducing it to a series of knowable themes, "whistleable ditties."While this may well have been true, what is gathered from the generality of popular musical discourse at the time is an expectation that the audience would have had a firsthand experience of a live performance of any given piece of music discussed. This implicit separation of the discursive musical imagination from the sound made by music is historically significant. What it suggests is a condition whereby, in the mid-1930s, a philosophical approach derived from the context of performances of live music, produced by critics, historians, and other interested commentators, was supplied as the means of listening to a broadcast music that was dissimilar in kind from the live music it purported to represent. Fractioned so, the discrepancy between what was heard and what was read may have produced the type of suspicious reading that I have been discussing here: one that may have allowed for contingent readings not only of Pevsner, but also of the very idea of the coyly knowing social condition of suburban modernity.

———————

A further form of musical reading accompanies this view. In one of those many widely read introductions to the appreciation of concert music aimed at radio audiences, J. H. Elliot, the Manchester-based critic, started a discussion of the role of color in music. He described musical color in terms of timbre, the specific characteristics of particularly played oboes, for instance, and suggested that one important use of color was to differentiate between different coherent strands of musical argument. Color has a substructural role, here. Elliot was someone who preferred to think of musical historical issues in terms of extension rather than in terms of well-behaved linear progress. Also, he differed subtly from Adorno in his views about the way that radio changed music. Where Adorno heard the combined overtones and other aggregate sonorities of live music as the technologically vulnerable acoustic matrix that both articulated and comprehended a manifold, Elliot presumed that it was inattention to argumentative consequence and the moments where arguments cut across each other that led to a floundering confusion in music. Shoddy color reproduction could fool attention and affect the perception of rhythm, harmony, and meter.

It is just at this overinterpreted point that Sibelius, specifically as an adept in the use of the hemiola, becomes significant to the historical interests of a British suburban listenership. Many took a hand in his invention. For Cecil Gray, author of the first book-length monograph on Sibelius, Sibelius's scores became the test of the probity of any performer in their ability to realize what could be found between notes. For Neville Cardus, in a sense Gray's nemesis, the texts represented the opposite: the creative and negotiative occasion for the meeting of the musical minds of author and working performer. Elliott thought him eminently diverse and capable of bearing twisted half-quotations in "a torrent of oratory." Ernest Newman, speaking of the illusory nature of musical progress, argued that the by then familiar accusation of "formlessness" came as a result of too ardent a critical search for conventionally expected symmetries. Donald Tovey, formally describing his Fifth Symphony and unable to resist the adjectives of mystery, spoke of a dominating rhythm composed of several different tunes. He also described a symphony that ends "with all the finality of a work that knew from the outset exactly when its last note was due."[17] In this complex figuration of an acoustic personalization of an epochal *zeitgeist* all nevertheless agreed on the fugitive rhythmic complexity of Sibelius's music.

Tovey's conceit of the subjectively self-conscious, comprehending nature of some symphonic music emerged as the stake in perhaps the most immediately related interpretative text to a broadcast performance of Sibelius's Fifth Symphony on the National Programme on a Saturday evening in 1936. Ernest Walker provided the program notes published in the *Listener*. "Rightly," he told his readers, "we plume ourselves" that Sibelius has been more warmly received in England than anywhere outside his native Finland. And, having established English musical culture as an accepting one, Walker went on to draw a picture of the weighty rhythmic foundations from which "something vital" would emerge. In all, he offered a version of the symphony where complex, deep-lain rhythmic interaction would eventuate in the precipitation of recognizable musical entities. For some of those concerned, Tovey and Walker included, Sibelius's symphony was an apparently simple thing that organizes itself entirely toward the final exposition and resolution of the massive "hammer theme" that dominates the concluding rotations of the final *allegro molto* section. In this, and as with Pevsner, Sibelius is conventionally accused of a singularized, concluding simplification. To reiterate: an acknowledgment is made of the rhythmic complexity of the work, but the idea of the simple, "whistleable" theme is allowed to predominate. There is a kind of conceptual dehiscence between what was likely to be radiophonically available in people's homes, and that which Ernest Newman had described concerning Sibelius, as what "the audience thought they were hearing," informed by the diversity of popular commentaries.

As with Pevsner's concluding conceit, Sibelius's whistleable, concluding theme is a curious affair, especially given all the critical emphasis on its teleological, expository nature. It has a famous last note, one that supplies a wilting and banal resolution. It might be thought dejected, or shy. It is one that is involved in a theatrically portentous-sounding statement of the main theme as a realization of the preceding, only partly formed material. Each note of the phrase is distinctly enunciated, and with each note the final tonically resolving note is deferred, again and again, until the likelihood of its being played, the very inevitability that demands its approach, becomes more attenuated and questionable. When it comes, it may seem a deeply disappointing, retrospectively autobiographic closure to an open, complicated, and suggestive argument. Its parenthesized self-consciousness is enough to prompt suspicion about the arbitrary and expedient nature of its terminating function. More than anything, it appears to articulate a dubiousness about the whole project of articulating the single, historical theme. But, already, there is suspicion of a different order here.

Rhythmically, Sibelius's music is particularly prone to suffering from a lack of the distinct separation of its components. It can sound murkily confused, or simply start to lose its hold on listeners' attentions, especially in that structural-argumentative way that Elliot had suggested as a preferred way to listen to radio music. All the critical talk of a thematic emergence from half-uttered musical materials may have seemed irrelevant or misleading to listening audience. A live broadcast, monitored by dutiful engineers, working with insufficient equipment, was heard on generally limited receivers by people who were attentively listening for rhythmic nuances that may never have materialized. All those features of the Sibelian historic (or national mythic) landscape, those complex hemiolic effects signaling inexorably tumbling rivers and sweeping topographies, may or may not have been available to suburban audiences for further symbolic interpretation. As important indicators of an evaluable, Sibelian sentence-sound, these effects may or may not have even been heard. It is impossible to be certain.

This idea of a phantastic, morphologizing sonority, one conjured in the popular critical narratives on the work and separated from the actual sound made by domestic radio receivers in thousands of suburban homes, is important. The material acoustic facts of Sibelius's hemiolas alone may be regarded as forms of theoretical and historical knowledge in their own right. Their possible durative and involving implications, Baroque or otherwise, could have supplied a popular, aural-philosophical model for interpreting Pevsner's architectural modernity, as well as paralleling the conditions of an always contingent reflection on a suburban sense of involvement in the main, perceived narratives of modernity.

While Adorno in the early 1940s was content with the idea of a suburban listenership unified in their stupefying subjection to the insufficiencies of broadcasting technology, he neglected both what people knew of music otherwise and what they found in the contexts of listening to music. Suspecting device at all resolving points, and augmenting this skill through engagements with the plotting malfeasance of crime writers, suburban listeners to Sibelius's Fifth Symphony in 1936 could exercise quite animated senses of uncertainty and suspicion about the absence of particular kinds of expected rhythmic material from the acoustic representation of the music supplied by radio. Or, they could simply consider falling in love. That opening up of an extradiegetic space of interpretative possibilities *between* a phantastic and a material sonority, each equally dreamed, equally noetic, is all that is needed now to suggest the agency of a suburban audience in these differing articulations of the *history* of cultural modernity—particularly in the terms of a recognition of one's conditional absence from or presence in it.

3.1   Detail of Ludwig Mies van der Rohe,
      proposal for a convention hall, 1952–54.
      Marbleized paper, cutout photographs
      (of roof truss model and Republican
      Party Convention) on composition
      board. Digital image © The Museum
      of Modern Art/Licensed by SCALA/Art
      Resource, New York.

3 Reveals: Glass Houses, Stones

Martov could not have failed to understand this. Nevertheless he clung
to the idea of compromise — the thing upon which his whole policy
always stands or falls. "We must put a stop to the bloodshed ... ,"
he begins again. "Those are only rumours!," voices call out. "It is not
only rumours that we hear," he answers. "If you come to the window
you will hear cannon-shots." This is undeniable. When the Congress
quiets down, shots are audible without going to the window.

—Leon Trotsky[1]

Here, modernist architecture's identification of the glass wall with the glass window helps bring Ludwig van Mies van der Rohe together with Walt Disney's Seven Dwarves in a study of thinly sliced rock. Experimental composers Cornelius Cardew and John Cage move from there to develop a role for the musical window in the figuration of revolutionary and reactionary labor.

1 Strange Labor, Surreal Sound

"Hi-ho! Hi-ho! An F note called—*hi*—followed by another, one octave above—*ho*. In a mine, where a million diamonds shine, this relationship represents the entire politics of Disneyesque prescriptions for a happy, bulliable, anarcho-syndicalist organization of skilled labor." Body clock awry, in a room in New York's Gramercy Park Hotel, I wake at 3:34 a.m. I have been shaken by a sound and by this animated thought of work and song, or something very like it. I peer, uncertain of quite how awake I am, and write down that thought. In a month's time I know the Gramercy will close for a total renovation at the Miesian-minimal hand of John Pawson. And, I add to my notes:

In the composition of architectural space, windows figure as crucially in *Snow White* as in Wittkower … and in composing a rhythmic ground of bells and picks and hammers, Disney's composer-librettists, Larry Morey and Frank Churchill, did not allow those legendarily labouring little German immigrants to sing the octave interval chordally; but only as a call and choral response, as an interpellation. This is not polyphonous. It is not *part* song, and the sentimental individuality of each toiling voice is preserved in the choir, isolatable and oppressable.

Improvisations prompted by sounds on waking, even when as mechanical and drearily professional as this, have a kind of tedious familiarity and aren't often taken seriously. They also form a staple of sensible avant-garde references in fine art practice. Salvador Dalí's 1944 painting *Dream Caused by the Flight of a Bee around a Pomegranate One Second before Awakening* is just one of many cultural stagings of instants of intensely inventive, revelatory poetic activity at the threshold of opened eyes. It is true that the creative-cognitive erotics involved in the processes of this instant of rousing (it is not a hailing) by a bee's busyness may be more sensed than usefully thought through in the roaring pornographies of Dalí's painting. His populist misprision of Freudian psychoanalytic tropes was likely meant to be comic.

Drawing disparately from an iconography of autographic materials, Dalí repeatedly attempted to provide a portal onto the odd workings of that florid instant of waking creative perception. As an intentional paradox, these strangenesses are highlighted against a continuing presence of a spatially consistent ground. That consistency may be figured as a limpid sky, his relationship to Gala as the organization of a febrile ardency, the uncanny emptiness of a Spanish landscape, or the laboring, well-formed monotony in the onomatopoeia of a bee's buzz. In each case, these grounding figures provide the unsettling plausibility of encounters portrayed.

————

In 1929, Walter Benjamin articulated the view that surrealism may be seen to have offered the precipitate of a certain form of existence, while withholding that existence itself.[2] By the transformation of conventionally understood, normalizing, transparently commonsensical space, surrealism produced a *location* for a revolutionary subjective politics, if not that revolution itself. This process of the production of disparate spaces that may be rendered connectable and consistent only through a radical upsetting of supposed bourgeois cultural-perceptual normalities is recognized as part of a continuing philosophical and political legacy of surrealism. Benjamin was clear that the role of the city itself was critical for the likes of Louis Aragon, Philippe Soupault, and Robert Desnos. Benjamin was also clear in his attitudes concerning the ways that surrealism included architecture in this confabulation of space. He pointed to the substance of surrealist montage, to the quoted mute witness of lightning conductors, gutters, verandas, and weathercocks. He also noted the distressing transparency of the glass house of Breton's *Nadja*, and suggested that there, as elsewhere, what is also found at the window-wall of that house is the potentially revolutionary figure of transparency itself.

That other critic of surrealism, Roger Caillois, was suspicious of the easy assimilability of acts of spatial disjunction to the repertoires of more conventional aesthetics. In resistance to such compossibility, he developed an exemplary methodological critique of the regular and transparently consistent grounds of inquiry. This was to see what else may be irrationally surrendered when the hysterical regularity of consistency is pressed uncompromisingly by reverie. His short book on the graphic textuality of micrometrically cut and polished gems, *The Writing of Stones*, is a case in point.[3] Flummoxed and fascinated by the pictures of faces, body parts, and landscapes that appear to be contained by cut minerals, and by the problem of how one could countenance their intentional existence, Caillois developed an argument. Revealing yet opaque to reason, the images found in stones cannot represent a work. They can only later occasion the improvisation of work.

Caillois's view has interesting parallels within modernist architectural fixations—especially those relating to the idea of ever more thinly sliced rock as an enigmatically improvised confusion of wall with window. Such equations were pursued into visual-spatial language by Ludwig Mies van der Rohe and, as Massimo

Cacciari has noticed, by Adolf Loos.[4] This confusion is part of the conventional language of modernist architecture. It is seen in the design, for instance, of the translucent window-walls of the reading room of the antique documents library at Yale University, by the firm of Skidmore, Owings, Merrill. Architecture is heavily invested in the enigmatization of the window-wall as an image of the condition of knowledge.

———————

Elsewhere, Caillois remarked on surrealist interest in figurative imagery:

Surrealist painters do depict human figures, but these are always engaged in baffling activities, in mysterious circumstances and settings: they seem to be illustrating an episode from some enigmatic adventure that no will ever know anything about.[5]

This observation on the dislocation of understandable action from the social grounds that provide that very understandability might also be a definition of the poetic articulation of the elusive meaning of the labor of the proletariat, both in the lumpen and the more noble, revolutionarily self-aware forms that have been lent to it.

2   The Temper of Transparency

Hi-ho! The sound that has just woken me came rising up through my open hotel window. Against the murmurous background of ventilation plant, creeping traffic, and the occasional chirrup of car horns and sparrows, it is an emergent sound, rather beautifully formed. A drunken growl, starting low and urgent, made of several different word-sounds, each almost monadic, without parts, slurring in their different directions, blasphemous, profane, angered, resentful; bleating betrayal, personal injustice, and fantastic retribution. It ascends loudly to crescendo. Then, gaining an elegant legato, in a broken and hesitant cadence of more discernibly defeated words, it comes to a breathless end; with a mineral coda of rolling glass—a bottle? The whole wail becomes speckled with other voices, those of deprecation. "Shut up, you! You, shut up!" These bring further refrains, each carrying with them, as a locating signature, both accent and the architectural resonances of the hotel rooms from which they originated.

It is a formally appealing cluster and reminds me of why I am here in New York. I am here for a musical archival reason, a reason that is also the mechanism that allowed me in the first place the possibility of appreciating the detailed musicality of this outpour, as if it were caught in some kind of aural vitrine. Carrying the symptom of my unsettled state of mind—a minidisk recorder and a microphone cheaper than I'd like—I am in Manhattan again to try to make a field recording, to attempt to capture a relationship, some trace of what an unnamed critic took to be an interruption from a window.

At the beginning of *Indeterminacy*, his long, anecdotally structured work that he recorded with David Tudor in 1959, John Cage says the following:

One day when the windows were open, Christian Wolff played one of his pieces at the piano. Sounds of traffic, boat horns, were heard not only during the silences in the music, but, being louder, were more easily heard than the piano sounds themselves. Afterward, someone asked Christian Wolff to play the piece again with the windows closed. Christian Wolff said he'd be glad to, but that it wasn't really necessary, since the sounds of the environment were in no sense an interruption of those of the music.

It takes Cage one minute to read, and it appears in the text of *Indeterminacy* only moments before he relates the more famous tale of his experience of the way the silence that he expected to hear in an anechoic chamber at Harvard University was drowned by the throbbing and tinkling of his vascular and nervous systems competing for his attention. This passage involving Wolff's window and his waspishness is endlessly ambiguous. *Indeterminacy* represents a curiously fraught mode of autobiography on Cage's part. Placing Cage at the center of the work, *Indeterminacy* is also designed to forestall romantic assumptions concerning tortured musical catharsis. Orienting itself in the direction of an essentially goalless musical practice and toward whatever else may be perceived when a sensibly described ambition is deleted, it is also an essay about perplexing and seemingly pointless though not fruitless labor. Another anecdote in *Indeterminacy* recounts a time when David Tudor went through a box of palm sugar, spice, and beans that had been delivered to him, all hopelessly mixed up and damaged. Over weeks, he separated each bean from each speck of spice and sugar. Eventually, the anecdote continues, Tudor offered to sort through a similarly damaged box that had been sent to Cage. Clearly, some pronounced technical brilliance of absorbtion, possibly valuable, had developed here. Cage refused of Tudor's offer. This refusal may—but only may—signify a desire to preserve such skill from the deleterious effects of its integration into an economy, no matter how esoteric that economy might be.

A favored reading of Cage's windowless, soundless, anechoic story seems to knowingly structure a disappointment of the desire to find the subjective treasure of a silence. It might work to ensure that silence be understood not as an object or condition, but as a perception. His story of Tudor's obsession figures an attempt to deny capitalism its capacity to assimilate forms of labor. And, his Wolff anecdote upsets the status and standing of the musical object and its audience, radically. Radically, that is, to the extent of offering the window and its transparency as something whose complete oracularity is on a par with that of architecture's understanding of the sliced stone window-wall.[6]

———

Over time, Cage came to know very well the problems of writing a life into a musical biography.[7] And his words on Wolff in the score of *Indeterminacy* bear the gestures of a reflective parable on the historical procedures of biography and autobiography. It is a reflection that resonates with Cage's understanding of the appreciation of the musical event. Within the subjective trivium of his musical understanding, the authoritative positions of audience, composer, and performer are equable and interchangeable. At times, all three positions can be occupied simultaneously in the same act of attention. Cage's spoken text represents a score around which interpretative invention is invited. His anecdote on Wolff and his window may impart sentiment, even nostalgia. It may also convey an aesthetic principle. It can, as well, offer historically specific documentary material. It includes the varied social connotations of Cage's own syntax. And, while that voice may also represent a further syntactical structure, it also offers that structure ambiguously, as a kind of narrative musical coherence and an acoustic substance on which to dwell and on which to map other, less immediate significances. Within this context, then, the text of the Wolff anecdote provides an opportunity to improvise on a set of themes about the liberties of self-electing position. This, for Cage, is how *Indeterminacy* is political. Via this unconstrained invention, what Cage perceived at the window as Christian Wolff's special form of uncompromising musical environmentalism constituted a musical revolt. But for Cage, music was social.

Provided by piano and tape, David Tudor's contribution to the 1959 recording of *Indeterminacy* comes as a series of epiphonemes, interjections, accompaniments, and musical remarks. They may appear random or ordered, depending, again from a Cagean vantage point, on how and where one chooses to situate oneself in relation to the proprieties of meaning that are established by the piece on each hearing or remembering.

However, at the moment where Cage speaks of Wolff, these standings become further questionable and, if possible, more deeply enigmatic. At times, when the text introduces the idea of interruption through the window from road and river, Tudor's abrupt piano chords and found and carefully manufactured electronic squawks start to take on something like the role and appearance of hoots. These appear illustrative, as something akin to the fleeting realism of Beethoven's pastoral illustration of lambs and lightning. While a mimetically considered urban landscaping isn't the only possible reading of Tudor's musical gestures, his provision of a set of phonemic equivalents to traffic and boat horns both supplies and deletes an important musical liberty. The inclusion of non-authored, intrusive, extraneous urban material into the field of attention organized and occasioned by Wolff's music has political implications in terms of its representative character. The accommodating Wolffean soundworld understands both boat horns and the sound of the piano to be mutual, to be compossibles. The allegorical implication might be the refutation of the right of interruptions to *maintain* an opaquely participatory, democratic standing *as a resistance*. For the violent cultural politics of America in the late 1950s and early 1960s, the preservation of the character of resistive, improvised interruption, as such, was important.

———

Wolff has confused matters about what was admitted by Cage's window. When asked once, he confessed to not remembering the specific occasion. He said something to the effect that it certainly did happen, and that he was happy to rely on Cage's story. It is impossible to reconstruct the extent to which this approximation of memory is a function of Wolff's thorough internalization of particular aspects of the history of the ethics and aesthetics of experimental music.

The performance that forms the basis of Cage's anecdote took place in his apartment on the sixth floor of a building on Monroe Street, since demolished, overlooking the East River. I am now here, having taken a circuitous route, from Gramercy Park. Having traipsed down Third Avenue, I have walked down past the sidewalk workshops of Manhattan's secondhand catering supplies industry, where shouts in mixed accents, the sounds of complaining stainless steel fittings, and the aerosol application of powerful grease-stripping detergents provide an acoustic stage for the redundancy and visual near obliteration of large shop windows, consigning them to a secondary and insufficient security function as walls, behind their rolled, aluminum shutters.

A substantial section of Monroe Street was demolished in the late 1950s to make way for the blocks of the East River Housing Cooperative (Corlear's Hook Houses) that stand there now. Picking my way around this elbow of the Lower East Side, I listen for boat horns. There are none. What else could realistically be expected? Also, and unsurprisingly perhaps, given the obsessional structure of this pilgrimage, I have forgotten my minidisk recorder. Improvising, on a mobile phone I dial my number in London, hold the phone at arm's length and leave myself an acoustic Post-It. On Grand Street now, at the point where it once intersected with Monroe, the precise site of Cage's apartment, I repeat the exercise, this time calling my cell phone. With the fleeting hope that this may be routed back to me via the monumentally windowless Long Landlines building owned by AT&T not so very far away, I head off with the intention of listening carefully to this recording: some cars, the rustle of my coat, sniffles, and the conversation of passersby whom I'd not noticed.

Nearly half a century later, not only does Cage's apartment window find itself translated by this foolish archival errand into the electrically backlit window of my phone, but a compelling memorial representation of labor also presents itself immediately. Samuel A. Spiegel Square, which sits at the lost intersection of Grand and Monroe, is a small and, with these November drafts, sometimes rather grim spot. It was named after the member of the New York State Assembly and Justice of the New York Supreme Court, who in 1962 sponsored an influential piece of state legislation. The Spiegel law was drafted with the intention of providing protection for tenants of the slum properties in this and other areas of New York. It allows tenants to withhold rent from any landlord who let his or her properties decay into an uninhabitable condition. From the legislation arose the frequently relied upon Spiegel Defense for disputes over legitimate arrears. Another benefit of the law came in the form of further financial goad for landlords to cooperate in the municipal development of social housing in the area. If ever a piece of legislation lent interruptive voice to the immigrant families in the Lower East Side in the years after the Second World War, it was this. Recently, however, its use has been overturned, or at least its penumbra of effect substantially eroded. In the case of Alexandria Rosario's refusal to pay her portion of rent to Notre Dame Leasing, giving the decrepitude of the building as her reason, citing broken windows, an adjudication declared that in instances where tenants were in receipt of state benefits of which payments for rent formed a part, the tenant had no right to withhold any payments, even their own portions, except in the (unlikely) event of the municipality making a similar withholding.

With this mind, I soon notice the memorial structure of the architecture of the East River Housing Cooperative itself. There is the street named after Abraham E. Kazan, for the political initiation of the cooperative housing movement that led to the proposal of the East River Housing development in 1950.[8] There is the Morris Sigman Building, the Erlich-Alter Building, the Benjamin Schlesinger Building, and the Morris Hillquit Building. The grounds of the Morris Hillquit Building cover the site of Cage's apartment. Each of these are named for key political figures in the history of socialism and in the history of the garment workers unions in this area.[9] This is a complex history of dispute and political representation, as well as less laudable intentions, which includes labor leaders like David Dubinsky as much as it does the more notorious actions of Lepke Buchalter.[10] In this sense the East River Houses produce one architectural image of labor in Manhattan.

3.2   View of the Morris Hillquit building. Part of the East River cooperative housing development, Lower East Side, Manhattan, Hermann Jessor & George W. Springsteen, 1955. It is the site of John Cage's since demolished apartment. Collection of the author.

3 Grumpy's Rent

The role of rent protection measures, both in the formation of the sociabilities of New York's cultural avant-gardes and the corollary contributions made more generally to the architectural imagination by the domestic desirability of airily lit and massively fenestrated loft spaces, is strictly the subject of other work. However, the historical substitution here of an urban monument to an already fading legal landmark for Cage's manifesto concerning misinformed interruption casts other archival light on the window as an architectural figure of speculative artistic labor.

Standing here, it is difficult to fully assimilate the enormous extent by which this area of New York has been transformed since Cage's time of occupancy in the early 1950s. Under the patronage of the Rockefeller family and the legitimation of successive mayorships, though with a great degree of self-excavated authority, Robert Moses, the character who so prompted Jane Jacobs in 1962 to a polemic of preservation, directed the clearance of hundreds of acres of New York's slums. He returned great blocks of land to a kind of planner's virginity in order to make way for thousands of units of social housing.[11] This process claimed Cage's tenement block and, with it, the residence that he'd had carved out from its upper floors.

Historically, this area of the Lower East Side also provided a source of powerfully convincing visual imagery for such socially crusading photographers as Jacob Riis and, later, Berenice Abbott. It is one of the locations at which the visual rhetorics of transparency, essential to the seductive plausibility of social realism, have come to be historically articulated. The photographic legacy of those engaged in social reform also provides a kind of memorial to the area and its inhabitants in the period up to the late 1940s. When that is coupled with a more currently pervasive interest in forms of cultural historical tourism in New York, however, it is clear that it is only through a complex tussle between memorial and amnesiac tendencies in the pragmatic ideologies that have sustained urban renewal policies in the city that any architectural image of those subjects of reform remains.[12]

Whatever it was that entered through a window into the soundworld depicted by John Cage, whatever else it was free to mean, it also indexed both dissipation and unrepeatability. It did this in ways that might be seen to embarrass the more anarchically liberated forms of musical enjoyment that he may have hoped for. The increasingly dilapidated desertion of Cage's immediate urban environment in the early 1950s was one source of its appeal to him and his circle. The population of German and Russian Jews who had left their own architecturally memorial marks (pickle barrels, synagogues, fire escapes) were happily making for more salubrious areas both off and around Manhattan. Patterns of wartime migration, combined with the concerted burden of historical restriction and exclusion acts that was only partially lifted by the quotas introduced by the 1945 Truman Directive on Displaced Persons, brought about a set of circumstances where a substantial proportion of tenement properties in the Lower East Side were empty by the early 1950s. Having only settled there in 1950, Cage and his fellow residents of the "Bozza Mansion" were served notice in 1953, and the building was dismantled. Cage contrived some of the most significant works of his oeuvre here, not only the tape piece *Williams Mix*, but *Imaginary Landscape IV* (for 12 radios), *4′33″*, *Perilous Night*, and *Music of Changes*. While one memorial function of this remembered building might be of cradling in time and place a period of politically, philosophically, and aesthetically influential avant-garde musical innovation, it is also apparent that the 1959 recorded version of *Indeterminacy*, Wolff's window, and Tudor's animation of it represents another, aesthetically ambiguous memorial function of this place.

The cajoling of certain elements of the anecdotal *mise en scène* of *Indeterminacy* into countenancing picturesque aspects of New York's urban history, and the suggestion of a mode of indexical socio-acoustic realism at work Tudor's contributions in the piece, is at odds with canonical views of Cage's experiment here. The way that Cage's window figures in the memorial and forgetful aspects of New York's urban history is that, in the instant of recognizing the possibility of an interruptive function for extraneous sound, *Indeterminacy* also recognizes that the productive aesthetic ambiguities that may be lent to that sound lay in the fact that, cartographically, from the well-formed, sociable hubbub of Lower East Side traffic, its window derived a presentiment that, increasingly as the pane in a vitrine, it had nothing left to collect, let alone comment upon. There was only a bafflement produced by the sound of the immigrant communities who had lived there which was being allowed to leak away.

————

Grumpy, as the image of unreasonably misanthropic bolshevism, and the only character of any real, if thoroughly compromised, sympathetic depth in the narrative of Disney's film, was of course enormously suspicious of Snow White's superciliously offered rent—a stew. "It's a witch's brew, I tell you." His peers, contrived as those little laboring others, with their genealogy-free, adjectival, idiosyncratic names, their curiously

LIVERPOOL JOHN MOORES UNIVERSITY
LEARNING SERVICES

infantilized bachelor ways, and their esoteric technical knowledge of the brilliance of stones, feel only anxiety when they return to see their house made neat as a pin by Snow White and her woodland friends. As the first hurdle to overcome in their successful integration into the consistent light of the renovated state that Snow White represents, this anxiety appears as an epiphenomenon of her work, as a concentrated metaphor of radical urban renewal. Sleepy is in heartbroken dismay at finding the carefully tended accretion of sugar in his tea mug gone. Spiders are irate at having their webs swept away by Stakhanovite enthusiasm, as are the mice when their homes are filled as idly improvised waste dumps. Each of these sensitivities is overridden by the disinterested, commandingly roughshod musical organization of labor: whistle while you work.

The window, variously conceived, has its own landmark functions in *Snow White*. The looking glass consulted by the Queen, the draped window from which she observes the Prince's courting of Snow White, and the cottage window through which she proffers a poisoned apple—each of these are respectively shaped as thresholds of emotional intent: envy, ardency, and malice. The forlorn melancholy that is ceremonially staged around Snow White's eventual vitrine, the woodland coffin of gold and glass wrought monumentally by the dwarves, represents a further improvised variation on the window's meaning.

As the formal convening of labor, and as a memorial identification with a monarchical state, the moments of the framing of Snow White's publicly perceived death rhyme architecturally and sentimentally with the moments when the dwarfs come home to see their house cleaned. Relating to this, Massimo Cacciari has indicated a small tradition of thought that has emerged around the meaning of the clean house, the modernist window, and the ideological function of its adventure, the glass wall. It is a simple and opaque fact of aesthetic detail that the reveals and mullions of the dwarves' cottage window home do not compare with the reflective spatial complexities allowed by those of Mies van der Rohe at Barcelona, the Lakeshore Drive apartments or the Farnsworth House. Cacciari's suggestion has been that a particular critical tradition, including observations made by Walter Benjamin and Giorgio Agamben, has noted the capacity of the glass wall. With its different sets of attendant aesthetic requirements, the glass wall is able, Cacciari considers, to render idiosyncratic the desire to mark and to leave traces of the occupation of modernist space, and in the process to annihilate any proliferation of singular aesthetics regarding the characterful accumulation of the motifs of significant cultural difference.

Cacciari's own position is more pessimistic than the tradition he details. It is not so much that Mies's glass wall is panoptically, narcissistically offended by collections, but that there is nothing left to collect. It is not that there should be nothing to add to this specific, replete, modernist architecture, but that there is nothing left that could be added. Mies's glass is unlike the authoritarian glass Walter Benjamin reserves for Jacobus Oud or Paul Scheerbart, and unlike the architecturally magical little gems proposed by Wenzel Hablik and other members of Bruno Taut's *Die Gläserne Kette* group. It certainly differs, too, from the unconfident glass that Philip Johnson employed in homage to Mies in his house at New Canaan. Here Elie Nadelman's sculptures and Poussin's *Death of Phocion* (an image with intriguing *étatist* parallels to the *Snow White* narrative: virtue, treason, death, mythic rebirth) provide a legend of Mediterranean vernacularity through which to interpret the building's surrounding woodland. This is a footnoting gesture rather alien to Miesian understanding of the reading of a prospect. Self-assuredly unlike each of these, Mies's glass is functionally transparent for Cacciari because, as utterly replete culturally as it is in its own view, it has nothing that it feels a requirement to collect, curate, convene, and explain.

Huddled behind a fir, quivering, uncertain, the dwarves find themselves alienated by their own home. The lights are lit. This is new, and through renewed windows they see that the traces of their own everyday have been scrubbed philanthropically away, the ciphers of their selves erased. The dwarves overcome their trepidation, and (after a cursory threat of mortal violence to her) reconcile both to Snow White's new regime of constant light and to the fact that it is merely monarchical tradition and not a white and ghostly modernity that haunts their house, after all. Their compliance takes the form of unthinking recourse to an ordinarily hospitable civility.

Not all were so biddable. Grumpy, who will not speak civilly, and the libidinously uncontainable Dopey, who will not even attempt to speak at all, both represent unformulated desires to maintain the resistively interruptive form of their critical support for their peers' democratic acquiescence to change. Such is the persuasively imperial, anthropological force of Snow White's requirement that it be demonstrated that she knows and is able to craft some happy, debt-laden memory of her downtown subjects, the dwarves feel compelled to perform. It is literally, again as if in a vitrine, that by their "Silly Song" the easily fluent improvisational accomplishment of Grumpy at the calliope and Dopey's percussion are revealed—even if, in a vain act of recuperation, it manifests as something of a war dance. *Snow White and the Seven Dwarves*, being what it is, in the denouements provided by Ted Sears, Otto Englander, Webb Smith, and other members of the regular story adaptation teams from the classic period of Disney cartooning, by the end everyone is on board and behaving, and the dwarves' folksy musical idiom is preserved as a charmingly inoffensive quaintness: a sheer skill, and culturally, essentially goalless.

However, the tiny, tremulous imperfections in the glacially crystalline, overenunciated vernacular clarity of Adriana Caselotti's voice in singing Snow White still hint at the refusals and disorder that Disney's narrative might otherwise appear keen to overwrite. This resistive character of improvisation at the site of the exercise of pronounced technical skill is a kind of performative politics of ornament—not always conscious or desired perhaps—that parallels what came, by 1972, to be seen as a darkened side of John Cage's conception of improvisation. In 1959, the reformulation of the kinds of traces of a life of labor that Cage and Tudor had Wolff's window figure had come to represent the ambiguous objects of avant-garde musical perception. By 1972, in Britain at least, these traces had come to be seen as the stake in a dispute over the political responsibilities of experimental music.

4 Not Everyone Loved John Cage

Eventually, British composer Cornelius Cardew fell out publicly with John Cage, over his approaches to the relationship of skilled labor to music making. For a time at least, though, he agreed with Cage about one thing: Beethoven. The German was, as Cage had put it earlier, "in error," and had an influence as "extensive as it has been lamentable."[13] In 1970, Cardew, the once adoring pupil of Stockhausen and advocate of Cagean principle, denounced Beethoven publicly on television: he was a "horrible monster on your back" who "permanently alters the structure of your mind."[14] These period remarks register the importance at the time of Beethoven as a touchstone of authoritarian villainy for many experimental musicians. Note the role of Wendy Carlos's refashioning of the Ninth Symphony in Stanley Kubrick's *A Clockwork Orange*, for instance. Not only this: regarding the supposed transparency of the score, these remarks are also register the powerful emotional and ethical significances that were attached to the governance of the conditions of improvisation as a means of interrupting narratives of good interpretative probity.

It may be said that it is a simple and opaque historical fact that no two performances from any Beethovenian score have been the same. The opening gestures of the *Waldstein* sonata, or the slow drunkenness of the second movement of the *Appassionata*, represent opportunities to articulate subjective difference, performance to performance. But, neither for Cage nor for Cardew could this ringing of changes be sufficient reparation for what they both saw as the institutionally debased condition of modern musical attention. The determination of tone, pitch, timbre, and other annotated features of common scoring and the approaches conventionally invited by this mode of notation, as far as Cage was concerned in 1958, does allow for a small portal at which varied perceptions of a musical object may be performed.[15] However, it is precisely the assumed function of the score in representing a vantage point on a preexisting musical object to be perceived and autobiographically nuanced, rather than as a mechanism for renewed perception itself, that structured Cage's views at this point.

In lectures given at Darmstadt in 1958, Cage proposed a view of a score in which the performer acts as "a contractor who, following an architect's blueprint, constructs a building."[16] By signifying the removal of the probities of musical performance inherited "from the body of European musical conventions," Cage's hope was that an improvisatory procedural approach could result in music that was not figured by institutionally communicated, rhetorical emotional poses. Cage prioritized the act of listening not as passive reception, but as interpretation. Cage was clear in his view that although his compositional procedures did produce new musical material, the modes of attention and perception, and the forestalling of the performer's conscious control of the sound produced, were the real points of issue. And, with this identification of a sound with a score, no matter what eventuates in the performance, a piece of music should not be paradigmatically representative of anything at all, let alone itself.[17] For each definitively "virgin" gesture in this a moment of free association for the liberated and distanciated listener is thereby sponsored.

It is indisputable that Cage's view of experimental probity was couched in an Emersonian cultural nationalism and against the image of the deadening hand of European tradition. But it was not for these reasons that Cornelius Cardew came to dispute Cage's anarchistic political findings.

5   Windows Are Not Transparent

We will return to the politics of improvisational and aleatory musical procedures in the 1970s shortly. But first, it is worth making a digression into the tropology of architectural glass, in order to imagine a complex ground for the interpretation of transparency.

Architectural history, as a whole, does not know what a window is—or at least it can't agree. Glass, reveals, mullions, and all the other material facts of the window, yes; but it can't agree on its meaning. It is true that the remarks Massimo Cacciari has made on Miesian glass question the ideological nature of its transparency. They do this as a kind of Heideggerian contribution to a broader, recognizable strain of critical architectural theory that has at its core a Marxist theory of the subject. However, in his approach, Cacciari is in no way representative of approaches to the window. For example, in his implication of musical theory with architectural theory in the Baroque period, Rudolf Wittkower identified a numerological and spiritual cosmogony at work in the conventional proportions of windows. The formal, external articulation of fenestration as argued for by Karl Friedrich Schinkel or by Étienne-Louis Boullée makes arguments about the calmness inspired by order, about the threateningly massive substantiality of walls, or the antique ratification of bourgeois conceptions of emergent democracy. More directly, the role of the window in framing a cultivated landscape, especially as argued for by Richard Payne Knight, was taken up by those aspects of thought relating to the picturesque that celebrated, in particular, the radicalism that became attached to the free association of ideas. For Le Corbusier, as rhetorically played evidence of the use of a load-bearing ferro-concrete frame, the aesthetic liberation of the facade was something articulated most vividly by the ribbon window. Pertinently, regarding this, in centrifugally propelling the subject to the peripheries of the house, as Beatriz Colomina has argued, the window functions to recall the cars and planes and fresh air that came to figure the Corbusian view of meritocratically mobile, modern, domestic sensibility. Lorenz Eitner, Karsten Harries, and Rosalind Krauss see the figures at Caspar David Friedrich's windows to be engrossed in kinds of romantic meditation on self and purpose. Walter Benjamin and Thorstein Veblen saw reflections in windows as critical instants in the seductions of capitalism. In unlikely ways, too, Roger Scruton and Robert Graves have found themselves in agreement over the moral communication of vernacular standards by the window. Mies, with characteristically oblique brevity, dismissed the window entirely with a question: "What is an opening?" For Jean Nouvel and his building for L'Institut du Monde Arabe in Paris, windows represent an institutional conflation of technically available means of self-regulation for the modern archive with a sentimental attachment to the particular light given by the screens of traditional buildings found in North African houses. Steen Eiler Rasmussen, speaking of English architectural values, argued that there was a purpose to the poor fit of English windows in encouraging a drafty counterpoint to the social comforts offered by the hearth. In other details of British housing design, Raymond Unwin and others insisted on the need for the admission of light to buildings, which led to legislations that a specific distance of seventy feet be between the back of one house and the front of another. That same seventy feet changed its meaning from health to one concerning privacy during the 1970s to guarantee the subjective freedom from being seen or overseeing, hearing or being overheard. And although, like others, Nikolaus Pevsner may be fairly accused of confusing the linearity of C. F. A. Voysey's medievalized windows with that of Le Corbusier's, he was elsewhere certain to ensure that the suppressed yearnings of Jane Austen's characters figured in the window-side architectural settings of her novels.[18]

———

Far from transparent, glass is in fact highly figured. In none of these instants just mentioned, and there are a remarkable number besides, is any single and cohering sense of a strictly apprehendable, transparently mean-ingless utilitarianism allowed to dominate as the window's functional meaning. The divergence of possible opinion concerning what might happen at the window suggests a terrain over which Cornelius Cardew and John Cage may have disagreed.

In appraising Cardew's reassessment of Cage, and the figure of the window in it, it is worth isolating one stream of thought from this pool of diversity. Beatriz Colomina, in her book *Privacy and Publicity*, has devel-oped a special kind of critique of the function of windows in Adolf Loos's domestic architecture. Promoting a view of the form of the detective novel as a paradigm for inquiries into the mechanisms of architectural space, Colomina's arguments highlight the peculiar opacity of Loos's exterior windows. She makes a case for the function of interior windows in framing, as if in a series of emptied vignettes, a paranoid and critical review of one's passage through his buildings. Cornelius Cardew returned surprisingly frequently to the image of the window in his critical writing on music. In the same ways that Colomina has suggested a poetics of forensic retrospection as a feature of Loos's interior windows, so too Cardew came eventually to value glass as a vehicle of valuable self-critical retrospection.

———

In 1972 Cardew publicly dismissed the entirety of his own previous work. He allowed the continued perfor-mance of some of these earlier works, works that he now thought decadent, on the condition that they were interpreted as essays in how *not* to do it. He took John Cage down with him. His article, "John Cage: Ghost or Monster," which eventually found its home in his polemical book *Stockhausen Serves Imperialism*, was first published to a broad audience in Britain in the *Listener* magazine in May of that year. This year was a politically tense moment in Britain. Cardew's article came as a critical, interpretive prelude to a concert of Cage's work to be broadcast by the BBC.[19]

The conceits of this text are various and manifold, but the image of the mirror is most compelling. One of the things the text itself mirrors is the formal repetitive structure of Cage's essay *Indeterminacy*. Cardew also expressed his approval of the attitudes contained in *Indeterminacy* regarding the liberation of the score. He was, however, unconvinced by the line taken by Cage concerning the enforced redundancy of the creativity of the performer. This was not in any sense a romantic conservatism. Moreover, though damning Cage in 1972, Cardew sustained Cage's argument about Stockhausen's music and his idea that its insistent scoring merely reified reactionary musical perceptions.

It is important to see that in reaching figurally for a mirror to see Cage, Cardew saw not himself but the Mao Tse-tung of the 1942 "Talks at the Yenan Forum on Literature and Art." Cardew found in Mao's procedures for self-assessment an image of paroxysmic Cagean sound as the figure of the surface dynamism of modern life that refuses to address tensions and contradictions underlying that surface. Cardew detected in Cage an untimely, counterrevolutionary obscurantism. For Cardew, Cage's work from 1958 absolutely required the input of dramatic emotive gestures, if it was to succeed aesthetically and avoid utter desiccation. His views, then, of Cage's procedures as simply throwing up acoustic material for passive contemplation also augured a critique of the control divested of musicians by Cage's music and a particular and crucial sociopolitical role for that deleted agency in improvised music.

In discrediting aspects of his own work, Cardew particularly focused on *Paragraph 1* of *The Great Learning* (1968–71). This piece, part of a much larger score, has become iconic of the traditions of British musical experiment, and comprises four main sections that include a "stones chorus" and the choral recitation of a section of Ezra Pound's translation of Confucius's *The Great Digest*. Structured as a tribute to the Scratch Orchestra, to Christian Wolff, to Terry Riley, and others, *The Great Learning* in part exploits the potential for varied interpretation of graphemes derived from Chinese pictograms.[20] The first two parts of the piece significantly attracted Cardew's criticism. His loss of temper with Pound is perhaps understandable, and Cardew put it clearly enough. Confucius was, he said, a reactionary thinker trying to "prop up a decadent and dying social system."[21] Confucius's political philosophy was, he thought, never put into practice and functioned "better in the ideological sphere as a means of deception." And in this sphere, Pound's translations drove Cardew to more incisive criticism. The second section of *Paragraph 1* requires its performers to read:

The Great Learning takes root in clarifying the way wherein the intelligence increases through the process of looking straight into one's own heart and acting on the result; it is rooted in watching with affection the way people grow, it is rooted in coming to rest, being at ease in perfect equity.

Cardew took this sentiment to indicate a social detachment that was characteristic of the totalitarian thought with which Pound is often associated. He specifically denounced Pound's phrase "watching with affection," and added the words "(as if through a window) the struggles of the people."[22] Cardew proposed both a more literal account of Confucius's lines and a more extensive revision for a promenade concert performance of *Paragraph 1*. This revision was more suggestive of an overtly materialist reflection, indicating a love for the broad masses of the people, and justice and equality as the highest good for all. Even this he eventually rejected each as reformist and unsound.[23]

It would be too easy to preserve Cardew's position, and extend his criticisms of the ideological content of the literary elements of his other works. However, this would dramatically reduce potentials for the interpretations of the pieces as they stood. The Stones Chorus component of *Paragraph 1* is a particular case in point. As a piece of composition it allows for the occasioning of unseizable, unrecordable aspects of collective cultural association in the formation of a type of transient community. It was written shortly before the founding of the Scratch Orchestra. The Scratch Orchestra was a changing collective of individuals with interests in experimental music and its possible politics. They were, proudly, not possessed of conventionally recognizable technical musical competences. In that functional sense, the Stones Chorus represents the form of a creative social amenity. It is inhabitable.

The scoring for this passage is apparently simple, though in fact shot through with tensions. Over the humming ground of a sustained chord on an electric organ, performers are required to play a series of quavers represented by what looks like common notation, but is in fact a formally graphic transliteration of the first few characters of the original Confucian text. The conductor marks five bars of three slow beats, irregularly spaced, and the performers tap stones together, individually as they think appropriate within the constraints of the temporality afforded by the conductor. The stems of the quavers are of different lengths. They supply an indication of the manner and degree of force with which the stones are struck. While this may register acoustically as an intensity of attack, it may equally evince only physically as a visible gesture, or in some other symbolic way. Fugitive aspects of the playing, then, may not be recordable acoustically, but in performance they may visually shape an aural apprehension. Other details of the performance of the Stones Chorus indicate that this is a passage of improvised translation of the text through creative labor. The syntax and accent of the performers inform the process.

What is produced is a clattering cadence that is anecdotal in character, and the archive of recordings of this shows a great deal of variation in how this may be achieved. Another feature worth noting of the performances of *Paragraph 1* and the Stones Chorus in particular is the role of glances between performers, indicating moments of readiness or intention.

The stones for the original performances of the work were chosen for their resonant qualities. Stories of rain-sodden pilgrimages across the hills of northern England in a barely roadworthy van to collect specifically

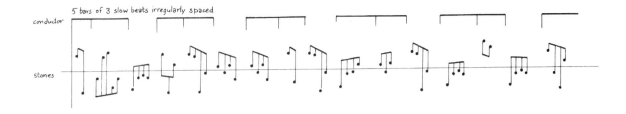

3.3   Cornelius Cardew, introductory scoring for "The Stones Chorus," *Paragraph 1* of *The Great Learning*. © Horace Cardew.

musical stones from a quarry make up part of the legend surrounding the work, and one struggles to keep the image of Disney's dwarves from one's mind.[24] The stones themselves, as the material of the romantic textures of regional vernacular architectures, supply both an indexical sense of imported place and an disconcerting confusion of the instruments of architecture with those of music making—and labor, and landscape, and a host of other connotations.

Moreover, the Stones Chorus sits in a context of observations made by Cardew as to the role of sociability. His book *Scratch Music*, along with a definition of what *scratch* represents, and a copy of the draft constitution of the Scratch Orchestra, also contains sketches for works, intended initially as improvisational, virtuosic solos. These were regarded as apprentice works, and each member of the Scratch Orchestra was required to maintain a book detailing his or her own lexicon of scratch procedures. Cardew's varied inventions in this vein are informative: "Tune a brook by moving the stones in it" or "If inside play the sounds from outside. If outside play the sounds from inside."[25] And, perhaps the most indicative of the social mien of scratch was this:

Take a mat and a cushion. Arrange instruments within easy reach for playing when lying down. Play, now and then, lying down. Go for a walk to see what others are doing, now and then. Maybe play an instrument while walking.[26]

Couple these with imprecations to "moan quietly and sadly, mouth open" or to "play specific sounds when people leave, enter, speak, look at you"; and it is easy to see why although, through these procedures, the whole of phenomenal experience may come to constitute a score upon which to improvise, the Cardew who read Mao might feel dissatisfied with this autobiographical portrait of a lazily decadent, passive, and sentimental aesthetic sociability: one too refined and esoteric to possess any urgently instrumental political purpose.

Cardew's transition of values, in his shift from avant-garde humanist aesthete to an explicitly committed political activist attuned to the worthiness of violence in political insurrection, speaks of its time. His desertion of these works does too. In what amounts to a philosophical migration from Wittgenstein to Mao, what Cardew had at his intellectual disposal for the interpretation of his own music making was the means of only an ideological-*textual* analysis, not a political-*textual* one. This may explain the way he so easily dismissed the entirety of *The Great Learning*, Stones Chorus and all.

Where that political-textural analysis did exist, however, was exactly in his musical practice. In his 1967 essay, "Toward an Ethic of Improvisation," Cardew detailed his hopes for the ways in which the social practice of improvisational music could resist both commodification and reification. As practices of association, the kinds of improvisation that he advocated were not, he thought, suitable "background for social intercourse." Because they lack precisely "logic" and regularity, because they vitally include agency and extraneity, he thought they were also resistant to any mode of contained commercial repeatability. In the same way as a glance from someone entering a room may have an effect on all details of a sound made by a scratch performer, and because Cardew thought it impossible to properly record music that is actually derived from the notational possibilities of the room in which it is made—"its size, shape, acoustical properties, even the view from the window"—the Cardew who had not as yet submitted to Mao was able to sense and condition a revolutionarily subjective location for spontaneity.[27]

Cardew and his collaborators, especially the Scratch Orchestra and the improvisation group AMM, have been able to demonstrate a resistive character for necessarily esoteric approaches to transparently indexical sounds—of stones tapped together, of carpets beaten, of piano strings scraped, of bottles rolled and rattled. These sounds are resistive not so much in the ways that they may be chased back archivally to social sources, but in the ways in which they occasion the development of spontaneous languages of association, prompts and gestures that are often as visual as they are aural. Cardew was concerned with the specific terms of the restitution of the accomplishments of the skilled performer, when confronted with a score and a desire to collaborate. This is something that is not available in Cage's account of Wolff's window. However, that value is seen and heard, perhaps surprisingly, in the way that, when she sings Snow White, Andrea Casselotti's voice sounds, but is not quite, operatic. And it is also seen in the way, improvising at the calliope, Grumpy prompts Bashful to overcome his reticence and get on with his silly song.

6 Mies

This political cartography of experimental music will return us to the absent architecture of New York, to its sounds, and to the figuration of the presence of its immigrants. As an exclamation, there is little that is as convincing as a hammer blow. Ludwig Mies van der Rohe is known for an aphoristic, hypercontracted language of virtuous architectural intent that is at once as precise and as opaque as his buildings. "Less is more," "Building is giving form to Reality," "The opposite happens," "I like cut crystal"—these are the elements of an epithetical repertoire, which appear to parallel the abrupt and profoundly resolved architectural character of his most recognizable buildings.

There are recognizable Miesian devices: the glass corner, the travertine core, the regular standard use of light fittings, the single span roof, the free-standing onyx wall, the movable wood partition. These represent not just a set of proprietorial signatures, nor even merely the roughly translated ubiquities of a broadly adopted commercial language of tower block design. More, they are the grammarial marks of a spatial liberty, one that is simultaneously caught up in a discussion of totalitarian control. Mies stands as an icon of the irresistible presence of European architectural tradition and conceptual inheritance in the American city. He is one of those Europeans who gained access to American citizenship via the political thinking that led to the Truman Directive mentioned earlier. Attitudes to Mies's benignity have been mixed. The perceived unreasonableness of his aesthetically led desire to insist upon a timetable for the opening and closing of specified domestic window dressings for the Lake Shore Drive apartments seems to signify that Mies did not make room for improvisation in the domestic or indeed any other sphere of life that intersected with his view of the correct inhabitation of his architecture.

During the 1950s, suspicions arose about Mies's intentions. These can be seen in the kind of Emersonian counter-Americanism that Cage exercised. In 1953, Elizabeth Gordon, writing on the plight of the unhappily litigious Edith Farnsworth over the house Mies designed for her suggested similar sentiments. So too did Frank Lloyd Wright with cattish, opportunistic dismissals of Miesian modernity. The more recent critical concerns about Mies's relationships with the Third Reich between 1933 and 1937 have added to this wariness. Even his own comments about why, in the end, he chose against facing potentially disgruntled clients and decided not to move into one of the apartments he'd designed on Lake Shore Drive, have contributed to the spelling out of the reputation of a modernist cultural producer who was at odds with the fraught modes of American artistic liberty.[28]

There may be no continuous and totalizing view of a homological or expressive totality of American culture that can hold Cage and Mies together. Cage abandoned the Lower East Side for the Stony Point experimental community established by his patron, the Bauhaus-inspired architect Paul Williams. But that won't be enough to suggest that in this he made a gesture toward specific, imported conditions of architectural modernity, and thus established the grounds of a rapprochement with Miesian architectural forms and cultural imperatives.[29] Even their stated and exercised interests in nature are too dissimilar for this. On the other hand, Mies's considerations of Manhattan, as evinced by the serried apartment blocks he proposed in 1953 for the Battery Park City development, may offer a fingertip's purchase on a terrain across which differences may be sensibly articulated.

This much is clear: Mies van der Rohe's attitudes to the meaning of glass changed over time. The refractive mysticism associated with his understanding of crystal in the early 1920s shifted by the 1930s to embrace a view of the glass curtain-wall as the figure of the fulfilment of frankly revealed constructional integrity in a way that historian critics like Sigfried Giedion and Nikolaus Pevsner could applaud as the determining feature

3.4   Ludwig Mies van der Rohe, proposal for
       the development of Battery Park City,
       Manhattan, 1957. © Hedrich Blessing/
       Chicago History Museum.

of architectural modernity.[30] By the early 1950s, speaking of the Farnsworth House, Mies had shifted again, noting that neutral colors should be used in his interiors in order to let the striking mutability of natural color have free rein at the window-wall.

However, in 1966 he gave an interview to Radio in the American Sector (RIAS). In it he offered a glimpse of a set of working manners that are further at odds with his other recorded statements. He revealed, anecdotally, that he had trained as a mason and stucco artist. He said that he used to, and still could, draw the spirals of a cartouche at will, without thinking about them. It is a bit like skating, he said, a skill and competence, redundant and esoteric in the face of the requirements of modernist aesthetics perhaps, but a somatic knowledge that a person never loses. When requested to sketch an account of his time in Germany in the mid-1930s, he obliged. He made a possibly allegorical point about technical handcrafts, traditions, and the cultures that maintained them, despite the purloining of their symbolism by the *Reichskammern*. He suggested these represented a reason why some Germans may not have wished to leave Germany and that this did not necessarily carry with it an implication of undue political sympathies.

In this enigmatically autobiographical context, he described a particular aural architectural moment—canonical, perhaps. The building of the German Pavilion at Barcelona in 1928 is an event wreathed in legend. Historian Wolf Tegethoff has claimed this legend to be a modular one.[31] Tegethoff has sought to discredit the popular theories that there exists in the proportions of the German Pavilion some barely suppressed numerology that mystically organizes the spaces contained and implied by the building, and which is guided by the proportions of the free-standing *onyx doré* wall at the center of the structure. He has also pointed out that in the monochromatic images of the Pavilion that circulated at the time and have continued to do so since, the green Tinos marble used to frame the pool, in which Georg Kolbe's classically figured sculpture *Alba* stands, is indistinguishable from the conifers that stood behind it. Tegethoff has also insisted that there is something of significance in this to an understanding of some aspects of modernist architectural space. One implication is that the Tinos marble wall functions here as a kind of baffling hesitation on the threshold of a condition of both transparency *and* mimetic representation, in almost precisely the same way that Cage's Wolff anecdote suggests. Glass gets strange here.

In his radio interview Mies added to these legends. He suggested a relationship between the technical accomplishment of skilled labor and the marshaling of a controlled and precise aural violence. In searching for a suitably large-sized block of *onyx doré* for the Barcelona building, Mies approached a familiar dealer. Mies was told of a suitable piece of stone, but was also told that it was spoken for and destined to make vases for a new ocean-going steamer. Mies said:

"Listen: let me see it," and they at once shouted: No, no, no, that can't be done, for heaven's sake you mustn't touch this marvellous piece. But I said just give me a hammer and I'll show you how we used to do that at home. So, reluctantly they brought a hammer, and they were curious whether I would want to chip away a corner. But no, I hit the block hard, right in the middle and off came a thin slab the size of my hand.[32]

This architectural epiphoneme, this resonant bang, producing an enigmatically transparent sliver of semiprecious decoration for one of the majestic icons in the ontogenesis of the international style, has implications for the discursive production of Miesian glass. Here, though, the stake becomes not transparency, nor even reflection—at least not in the ways that the conventional approaches to the poetic figuration of the chaos of urban modernity may prefer it. Rather, we find a notion of a dissonant perception of improvised strangeness from the interior of the building.

The device of the glass corner is one that Mies preferred in in later years. It involves the handling of conjunctions of glass with glass, rather than with brick or any other constructional material. It also involves keeping all structural supporting elements as far away as possible from this corner. One of the phenomenal perceptions of Miesian glass in this device is that, from the interior of the building a viewing subject may catch a glimpse of himself in the repeating order of the building's own consistent internal organization. Moreover, at the glass corner, these reflections are seen to repeat and to project themselves out onto whichever urban or pastoral landscape the building provides prospects upon. This material poetic extensiveness of the building's regularity certainly does not overtly function as an aspect of the applied decoration of Mies's buildings from the period of the mid-1950s. Quite unlike the facadist rhetoric of his applied steelwork in his designs for the Illinois Institute of Technology campus, for example, this projected reflection is quite unannounced as an architectural feature. It is one perhaps captured only by photographers with the sensitivities of, for example, Guido Guidi. It is also one that, in exercising the simultaneous opacity and transparency of windows, suggests a figure in the glass.

The countervailing pressures of this projected order over the inchoate appeals of modern urban living have perhaps special resonances in New York. Mies's projection repeats the fantasy for the grid structure of Manhattan's streets and avenues that was becoming so important to the poetry and logic of Robert Moses's reconstruction of Lower Manhattan during this same time. And, there is no lack of images of Mies pulling contemplatively on a cigar and observing the nuances of the views afforded by his tall buildings from these glass corners.

3.5   Ludwig Mies van der Rohe, 880
      Lakeshore Drive, Chicago. Photograph:
      Guido Guidi. Courtesy Canadian Centre
      for Architecture. © Guido Guidi (Kate
      MacGarry, London, and Alessandro
      DeMarch, Milan).

Speaking in 1966, Mies may have been reaching for an aural-architectural image of technical facility in an allegorical manner. The anecdote is a dual one, relating a quite redundant knowledge of the behavior of particular types of stone in the circumstances of controlled violence, a skill honed by practiced labor, grounded in aesthetic conventions apparently both alien and baffling to the requirements of modernist probity. It amusingly overcomes a stone dealer's reticence and it improvises an enigmatic parable to the internationally ordering interests of America, as represented by European radio. But, it also reflects on a retrospective thought about the interpretation of Mies's career-long confusion of wall with window, about the guiding of a principle of investigative improvisation at the point of decisions about order or chaos, insistence and liberty, as well as the acceptance of European architectural sentiment in America.

Moreover, his words here may be thought to act to preserve as multivalent and resistive the various and discrepant images of busy labor vested in the changing landscape of the Lower East Side. As such, a memorial function is revealed in what may have been dreamed in his proposals for the Battery Park City project. His other unbuilt projects of the time seem to amplify this thematic of unruliness in improvised interpretation of what the window leaves. The 50 foot × 50 foot house project (1950–52), with its anxiety-inspiring placement of supporting elements, acts as a fully resolved essay on the use of the glass corner. As a ventured mass-housing solution, it makes the delightfully outrageous suggestion of a crystalline urban and suburban landscape upon which to dwell, one comprising the baffling disorder and asymmetries of others' lives—something rather at odds with the seclusions of Edith Farnsworth's home. More importantly, at this point, his jaw-slackening designs for a convention hall in Chicago, especially as seen by the photomontagic images produced to convince of its plausibility, show possibilities for association of varied kinds, whether political, celebratory, or commercial, with the improvisatory implications of the figured stone wall-windows as a legend to the hubbubs convened, with the Stars and Stripes leaping forward. In 1955, for Mies, the designer of the monument to Rosa Luxemburg and Karl Liebknecht, a monument to the violent relationship of the wall to the German working classes, these photomontages represent a flag-swathed reference to the isolationism of Robert Taft and Dwight D. Eisenhower. This image of the noise and unruliness of the American political convention possibly represents an innocent protestation about the awkward meaning of labor in Mies's past.

At the beginning of this essay, I spoke briefly of the importance of Roger Caillois in attitudes toward the aesthetics of animating the definitively intractable. In *The Writing of Stones*, Caillois attaches great importance to the lost interest in certain Florentine marbles, for the resemblances they offered to aspects of the external

3.6   Ludwig Mies van der Rohe,
       50 foot × 50 foot house project,
       1950–52. © Canadian Centre for
       Architecture, Montréal/DACS.

world, and particularly to dramatically mysterious landscapes. During the early part of the seventeenth century, these stones came to be valued, he says, as the subjectile of paintings. They suggested to artists and collectors the moment for the improvisatory depiction, with a few added visual remarks, of episodes from mythic tales of labor and endeavor. But, as a form of the interpretation and worth of stones as a field of representation or as a clear portal onto other worlds, this is an entirely unsatisfactory view for Caillois's purposes. Rather, the shimmering *significance* of stones is what interested him, the "extraordinary combination of signs which have no meaning but which are swiftly given a meaning that the ensnared imagination finds it hard to withhold."[33] Often, and in keeping with the peculiar social and erotic imagination of his intellectual peers, Caillois finds in American jaspers "an eye devoid of lid or lashes, or an empty socket with the freshly removed orb dangling like a wet rag or oyster from its shell," or gnarled phalluses and kneecaps and knuckle bones and glistening intestinal worms, rumbling innards, excited vulvas, and striated tendons. Caillois's figured windows here may not be compossible with those of Mies or Disney or Cage or Cardew. Equally, however, and as a further mineral coda that may provide a kind of terrain for these to speak to each other across a rethought field of musical and architectural modernity, Caillois could find in the labor of a forgotten Florentine artisan, sawing slice by slice through a limestone "a hallucination of palaces, ferns, birds and Lilliputians."[34] Hi-ho!

3.7   Ludwig Mies van der Rohe, proposal
       for a convention hall, 1952–54. Marble-
       ized paper, cutout photographs (of roof
       truss model and Republican Party
       Convention) on composition board.
       Digital image © The Museum of Modern
       Art/Licensed by SCALA/Art Resource,
       New York.

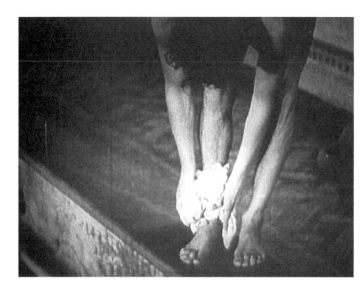

4.1 Jean Taris dries his feet at poolside in *Taris* (dir. Jean Vigo, 1931). © Gaumont.

4.2 Burt Lancaster demonstrates the architecture of Ned Merrill in *The Swimmer* (dir. Frank Perry, 1968). © Sony Pictures.

4 Splash

To the fleeting water speak: I am

—Rainer Maria Rilke[1]

Here we see Frank Lloyd Wright, Jean Vigo, Owen Williams, and others come together to produce something like a history of the sound of water and its capacity to speak with architecture, medically, existentially, ideologically.

———

The Olympic Games in Seoul. Mid-afternoon, September 19, 1988. The audience fell to silent expectation. Then, Greg Louganis strode out to end of the three-meter board of the Jamsil Stadium diving pool, sprang once and leapt upward. His aim was to execute a reverse 2½-somersault pike and qualify for the final in the event. On the way back down to the water, however, he cracked his head on the board and landed inert. As one, the audience gasped. Moments later, he emerged from the pool clutching his wounded scalp. It required five stitches.

4.3 American athlete Greg Louganis dashes his head against the three-meter diving board at the Seoul Olympics, 1988. © Empics.

Greg Louganis is in all ways an important sports figure. A poet in the air and water, he is winner of all accolades in his field and a significant touchstone for liberal commentary on social changes in athletic culture. He has a degree in drama. He is also one of those artist-athletes blessed with a capacity to bring his audience to a point of near-religious reverie before his competitive performances. As a prelude to any dive, he would stand, making himself the absolute center of attention, then place his hands together and bow his head. It was as if the audience went with him to another world of absorption into what was about to unfold.

As a diver, Louganis also articulated a kind of personal and mystical relationship with the water. Speaking of the instant immediately after completing another successful dive, he said, "That brief time underwater seemed a welcome relief from the events of the past few months." He added, "For much of my career, that time underwater had been a peaceful respite, a kind of friend."[2]

Water is the incomprehensible other of architecture. The relationships between the two always connote some mysteriousness. Louganis's memoir shows that water, as a species of architecture *autre*, can physically and subjectively offer quiet, equilibrium, support. It can even offer friendship. By hitting his head on the board and, in so doing, making such a contribution to the canon of aural-architectural objects in the twentieth century, Louganis also provided a moment of intersection for several discourses that make water and architecture cultural and complex. We have a moment here in Seoul, an acoustically indexed moment when Greg Louganis stopped *diving* and started *falling*. The aim of this chapter is to work out how that abrupt change in the demeanor of his flight altered what it was that he then fell into.

So, to work out the meaning of the sound of Louganis's collision, what I'd like to do is to work back to that moment in Seoul and figure a prehistory of a particular version of how architecture is with water. This will involve the formation of an appropriately rather syncopated and staccato tropology of poetic gestures toward water that involve a number of different notions. Principally these are to do with nature, health, hygiene and pharmacology. But, they also broach profound narcissisms and the formation of proper modes of family domesticity in relationship to Modernist institutional architecture. We'll also see details of peoples' sex lives, or at least State and public interest in them and the pool as something like an otherworldly heterotopia. This amounts to a conditional disavowal of the knowledge of the pedestrian flaneur, that is to say the conventional the subject of the architectural promenade, to be replaced by the more difficult self-knowledge of the swimmer.

1 Species of Water

Figurally, architecturally, water is nothing like inert. Its fluid and moody alterity animates some of the most iconic of architectural images. This can be seen in that way water reflects on and sharpens the brute and immobile bearing of architecture's structural qualities. This reflective capacity also means that water is able to bring an otherworldliness to architecture. The different kinds of spiritual hedonism that seem to be captured in the monumental tranquility of Richard Neutra's pool at Palm Springs, or Tadao Ando's pond at Tomamu, are good examples of this conventional architectural language and its way of putting various states of leisurely and contemplative being (figures 4.4 and 4.5). Both Neutra's pool and Ando's pond supply, as an architectural function, types of hushed ritual scenarios, opportunities for escape and momentary subjective suspension.

4.4   Julius Shulman's iconic photograph of
       Richard Neutra's Kauffman residence,
       Palm Springs, California, 1949. Photo-
       graph courtesy © Shulman Photography
       Archive, Getty Research Institute.

From drip to deluge, however, water has its own gently graded, aural vocabulary, one that can hardly be said to have been left unexploited by architecture. Water, and the spattering sound of it, even haunts the photographic manufacture of the modernity of some buildings via the practice of foreground wetting that was developed by Julius Schulman in the 1940s.[3] But, perhaps the most iconic exercise of modern architectural

4.5   Tadao Ando, Church on the Water,
      Tomamu, Japan, 1985–1988. Photograph
      courtesy Tadao Ando Architect
      & Associates.

4.6　Frank Lloyd Wright, Fallingwater,
Bear Run, Pennsylvania, 1935.
Photograph © Scott Frances/Esto.

water appears with the pantheistic and sentimental spirituality that Frank Lloyd Wright figured with his clever conceit at Fallingwater (1939). Fallingwater relies for its visual architectural effects on its mimicking of the features of its local geological environment. As such, it is like any other Frank Lloyd Wright building. However, acoustically, it is equally reliant upon the constant, aleatory gurglings of the Bear Run rivulet that passes under its cantilevered balconies. There is a certain arch staginess to the "found" character of the stream that Wright exploits here as an acoustic mechanism for living well. That theatricality hasn't checked architectural historians in their insistence that the tiny, ventriloquizing susurration made by the tinkling of water casts the building as a site of permanent and meditatively authentic reverie. It has become a stale aural image, and, tied to nationalist ideologies, it evokes taxidermy.

In 1956, during a singular storm, while much-needed renovations were underway to repair the effects of subsidence before the building was entrusted to the public, Fallingwater itself became a noisy cataract as surface runoff flooded through it. In a sense, identifying the building as fully as possible with its surroundings, this flood realized Wright's ambitions for an absolute integration of domestic architecture with its environment, naturalizing it. Although furniture and fittings suffered, the fabric of the building itself remained quite resilient to the stream that pattered uninvited down its stairways. Frank Lloyd Wright had no difficulties producing buildings that were tolerant of such events. As any architect might, especially one engaged in such adventurous constructional work, Wright wrote about the handling of such compromising pollutants and detailed the proper specification of valleys, flashings, and other rainwater goods. Few architects, however, continued to return to his themes of infiltration and corruption with such apparently lascivious relish. It is not "the deluge of water in a storm that hurts any building," Wright wrote then. Rather, it is the pernicious "ooze and drip of dirty water" that should guide architectural vigilance.[4]

Wright's fascination with waterborne impurity here is simply a variation on the mythic themes of cleansing, baptism, and transfiguration for water. This is where the otherworldliness of architecture's relationship with water begins to take shape. There are plenty of powerful examples to draw on. St. Angela of Foligno gave an account of a hospital visit where, as a kind of Communion, she sipped the water in which she had just washed the hands of a leper, euphorically catching a waferlike scab in her throat as she did. Hers is a description that makes clear that not every discursive approach to waterborne pollution disdains all redemption: sometimes the pollution of water is the act of redemption itself. Tom, the creation of novelist Charles Kingsley, is a young sweep who, on falling into an unnamed Yorkshire river, is cleansed, transmogrified, and then lives on as an Evangeline waterbaby to receive his Christian moral education before returning to human form. Tom allowed Kingsley to call attention to Mrs. Doasyouwouldbedoneby's views on the self-electing condition of virtue: "Those that wish to be clean, clean they will be; and those that wish to be foul, foul they will be. Remember."[5] Kingsley wasn't alone in playing out a popular soteriology of water through the development of purposefully wrought conceptual personae. In a more counterintuitive way, perhaps, Richard Wagner's Alberich finds transformation and a kind of liberation as he flicks aside Wagner's hateful notion of love (which isn't Alberich's anyway), and dives into the river to seize the Rheingold and everything that comes with it. This little set of incidents does a good job of introducing water as a locus of gravity for questions about virtue and good public moral conduct.

Perhaps the most telling figure in this tropology of transformations is Jean, as played by Jean Dasté in the role of the youthful and troubled barge skipper, in Jean Vigo's romantic comedy *L'Atalante* (1934). Jean, initially carefree and having just set off on a working honeymoon cruise with his new bride Juliette, suffers a kind of calenture. He becomes morose, and his actions erratic and unpredictable. His old friend, first mate and audiophile Père Jules, is worried by his behavior. Suspecting its cause, he relates a fragment of maritime lore to Jean: when your head is underwater, if you open your eyes you will see your one true love. Experimenting by first sticking his head in a bucket, Jean then plunges vigorously from his ponderous boat into the Seine. With his eyes wide open he enters another world. At that moment of fantastic equilibrium that is afforded by a physiological shock of entry, and the crashing aural violence and muted, bubbling quiet that follows, a screen appears before him. Upon it the image of his own unknown desire is projected. Fortunately for the narrative of the film, both image and desire coincide in their forms with that of Juliette.

L'Atalante is one of a pair of important essays on life underwater made by Jean Vigo. The other is his short documentary *Taris* (1931) on the French Olympic swimmer, yet another Jean, Jean Taris. Vigo's other Jean plunges in a more accomplished fashion. As a film, *Taris* is apparently educative in intent, but it is not simple to detect quite what is being put forward as correct behavior by the film. It isn't just a formally instructive swimming lesson. Certainly Jean Taris's charmingly competent manner, both in the water and out as national and Olympic hero, is easy to see, as he demonstrates a series of strokes and other techniques both poolside and from the starting block. The anarchically youthful appeal of the film appears to treasure something of an Epicurean delight in aquatic deportment, the inhabitation of another medium and the observation of the ebullient precocities of another, knowledgeable body. The film's unabashed enthusiasm appears in different

4.7 Skipper, Jean, prepares to dive from his boat into the Seine. Still from *L'Atalante* (dir. Jean Vigo, 1934). © Gaumont.

4.8 Jean sees his Juliette. Still from *L'Atalante* (dir. Jean Vigo, 1934). © Gaumont.

ways. In part, it is carried by Vigo's study of the common aural currency of the pool when used by athletes for innocent competitive pleasures. The competitions themselves are given as competitions as much with oneself as with others and also as a means of association. It is as if the pleasures obtained by competition could never be as socially and aesthetically pointless as they might publicly pretend.

2 Frank's Expression

The sound of *Taris* is quite distracting. Partly, this is because of a technical want in the means of sound recording that Vigo had at his disposal. Nevertheless, it is exactly by this acoustic approximateness that Vigo's sound-plan achieves what it does. It is never as interesting as the acoustic world manifest in Carl Theodor Dreyer's *Vampyr* (1928), a film that Raymond Durgnat first called attention to as an architectural essay during the 1960s.[6] *Taris* nevertheless manages to articulate the acoustic and architectural status of water in a way that

4.9a With a visual parallel to the acoustics of the film, the outline of Jean Taris's body blurs as he demonstrates his aggressive, arm-y style of front crawl.

4.9b Vigo introduces a theme of play to the somatics of his film. Stills from *Taris* (dir. Jean Vigo, 1931). © Gaumont.

is both special and pertinent. There are good reasons to compare Dreyer's earlier film with *Taris*. Its spoken dialogue is, like Vigo's, also carried by the insufficiencies of its recording medium. From moment to moment the repetitions, the blurrings and half-heard sentences, the pre-echoes of the soundtrack (which themselves make for the chillingly seductive aural mystery of *Vampyr*), are supported by a continuum of crackles, hisses, hops, and jumps. These appear in themselves both *at* and *as* the point of juxtaposition of each edited passage as it brings the differing characteristics of the particular sound recording media used to an equitable aural surface. The effects are both loud and unavoidable. It sounds as if the distorted and fragmentary, always recognizable yet not always decipherable linguistic phonemes present in the film are floating in a kind of disfiguring and dislimning acoustic medium.

This cinematic effect is paralleled in literature in well-known ways. In the section of *Ulysses* that meditates on the techniques and meanings of succession and juxtaposition in language, James Joyce has Stephen

Dedalus walking home alone, boozed up, thinking of himself and others. The phonemic elements of Joyce's passagework here start to identify Dedalus's drunken words and thoughts with waves and shingle, "seesoo, hrss, rsseeiss, ooos." In making this sibilance a medium that connects space with a disjunctive and associative subjectivity Joyce reveals a pharmacology at work, another register of shifted sensibility. Unlike Jean Taris, who is intoxicated with natural vigor, Stephen Dedalus is full of stout.

What Joyce provides here is a literary model for an etymologically speculative dialectic—a deconstruction—in which the origins of sounds become confused and transvocal by the sensibility of the subject in which they appear. The speaking subject that Joyce has at the edge of the waves at this point in *Ulysses* is neither the sea, nor the booze, nor Dedalus. Rather, the subject is a medium where tiny alluded archival details appear to relate to each other through the condition of their very approximation to recognizable speech. Joyce often constructed moments of narrative identity by playing with that acoustic near-similarity. Taken together, the soundtracks of Vigo's two films provide a cinematic version of that Joycean attempt to "deline the mare," and they do so to the extent that they comprise a handbook on techniques of identification both in and through water.[7]

———

So with this impression of what happens with the approximate, textural, and textual nature of sound, let's return again to Frank Lloyd Wright. Wright was a prolific writer. More sober than Joyce or even Vigo perhaps, he was not averse to similar such concrete experiments in glossomorphosis. His short essay on the monumental significance of Gertrude Stein, "The Man Who Was Gertrude," reveals a fallibility. In this surprising piece of writing from 1934, a passage plays repeatedly on the word *simple* to find a fluid commonality of sibilances with which to sustain an emerging prose space.

As anybody can see simples of such simplicity are unsimple to the too simple and are too simple to the unsimple so that simplicity goes off outside the inside of the too simple whereas off goes the simple from inside the outside of the perfectly simple.[8]

It goes on, formally. An ironic effect of its repetition, however, is to fix ever more fully the outline integrity of the word *simple* and the meanings secured by its aural boundaries. Wright's usage does not appear iterative; the word doesn't appear to change value at each repeat.[9] His usage consolidates. It is protective, territorial, isolationist. However, failures here in a piece of idly experimental writing (it succeeds in other ways) simply throw into relief Wright's prowess in refuting the conventionally assumed boundaries of architectural space. The aggregate susurration reached for by this piece of writing aims to rhyme with his use of the suffusing aurality of moving water at Fallingwater, which is itself paralleled by such signature visual architectural devices as allowing the materials of exterior walls to guide into an interior just as freely as materials from interior walls suggest direction out. Wright's domestic history was horrifying, and doorways figured terribly in that. But Wright also saw the varying topologies produced at points of spatial transition from outside to in as vitally democratic in effecting the new American domestic architecture that he was in the process of articulating. He went so far as to understand the harmonious fluidity of the central kitchen/hearth/living-room/dining-room nexus as a key emblem of the freedom of the individual as an individual. This imaging of different registers of suffusion—aural, literary, architectural—is an important aspect of his domestic polemic.

It is also significant that he saw "international style" gestures toward the discrete classification of architectural spaces as both unseemly and as somehow the emasculate offspring of his own "organic" architecture. His text on Stein works toward criticizing the sophisticated, cosmopolitan principle of simplicity in all things that she advocated. For Wright, the idiom of simplicity in design that she proposed is no more than an effete artifice, a construction, a fancy cosmetic. As far as he was concerned, the issue of simplicity should be left to an artist who is, as he says of himself, more genuinely and directly attuned to the "fine integrities of the more livable and gracious human simplicities."[10] The sibilance that Wright plays with appears as a critical vehicle. That remark on livability is a jibe aimed directly at Stein's domestic arrangements. It is one that, depending perhaps on how much Wright knew or was prepared to admit of the adventurous bohemian milieus that sustained Stein's aesthetics, may be thought to open onto a threshold of homophobic misogyny—onto something he might consider domestically polluting. Aiming to make an authentically simple sound, Wright found that he could not. For, in this sibilance, Stein, or at least his impression of her complex modernity, appears as sensuously lisping: the intellectual other to his American domestic decencies.

As with Joyce and Wright, the dialogue (and by extension its subjects) found in Carl Dreyer and Jean Vigo's early films is never disentangled from the noises of its support: that is also to say that it is never free of what may be regarded as its disfiguring pollutants. The dialogue of Dreyer's *Vampyr* is always, in part, constituted by differently characterized hisses. As the object of recording, it is the dialogue itself that provokes these incidentals. In this way, occluded and fragmentary as it is, the dialogue of *Vampyr* appears acoustically as a kind of desirable glitter on a darker, more rhythmically impressive and unwarranted surface. Similarly, Jean Vigo's only partial apprehension of the sound of regularly plashing water as Jean Taris moves through it augments a visually effected image of Jean Taris's identification with the medium in which he appears most lively. At points, both visually and acoustically, Taris becomes indistinguishable from the pool. At those moments, his less than elegant terrestrial habit is eclipsed by what the soundtrack and celluloid collude to figure as an abstract image of his easiness with the water.

4.10 Vigo starts to explore the luminous effects of pool water. Still from *Taris* (dir. Jean Vigo, 1931). © Gaumont.

Vigo's lecture about this schoolboys' hero, in its romanticism—and it's not criticizable for this—becomes one on becoming. This is not solely as a function of autobiographical identification on Vigo's part, necessarily. The disciplinary joys of lapping, the coordination of elbows with heels, the counting, the best management of opportunities to breathe, the dwelling on the universality of the pursuit, the immense status of other swimmers in their technical accomplishments and affinities, the ability to hear the diagnostic commentary of the water, and the ordering of all this to enjoy the forgetting of the pedestrian body: these are folded into the resistive spiritualities of Vigo's documentary and represent the film's amiably meritocratic view of exercise.

Taris may be regarded as an essay that senses the political virtues of a creative-cognitive sensuousness in exactly the way Frank Lloyd Wright could not. *Taris* was made at the time of another of the periodic paroxysms of interest in the control of syphilis, not just in France but in Europe and America. Looking back at the film's circumstance, a possible reading of it is in terms of discourses on public sexual health. At the same time, in its playful dignification and undoing of the somatic boundaries both of the body and of instructional manners at the poolside, *Taris* also questions capitalist requirements for specialized bodies when they appear in such specialized architectural forms as swimming pools. But it is worth noting too that Vigo made his films at exactly the time that particular techniques of swimming were being specified and detailed as standard Olympic forms. This tendency toward the aesthetic regulation of competitive performance is also recognized in Vigo's film. The poolside is different from the pool. The place of the audience, of judgment and administration, the poolside appears cold, stark, and architecturally austere. It appears as a platform for momentary quietism, with the swimmer held for inspection in statuary poise. The poolside is offered also as a permanent, unyielding, and marshalling feature: uncompromising, assertive, punishing of every frail failure of confidence or overconfidence. As a somatically figured threshold to the aural life of water, the poolside presents here as precisely the *other* of the underwater antics that Jean Taris's swimming presents as unpoliceable modes of ludically experimental, corporeal cognition.

3   Organic Architecture

Taris, as a utopian aural-architectural essay on athletic knowledge, nevertheless does not—perhaps cannot—deny the regimenting effect of the pool. In a sense, the image of Jean Taris's physical freedom appears only as a syncopated accompaniment to such restriction. It is true that prohibitions in such municipal environments do often appear as tempting invitations. Smoking, running, ducking, diving, bombing, petting: this is a familiar list of transgressions. At the same time, of course, they all read seductively as a menu of possibilities.

———————

The English Sports Council was established in the 1970s in response to a similar kind of resurgence of interest in the health benefits of organized physical activity that provides the context to Vigo's film. The council had an

4.11 Taris beckons. Still from *Taris* (dir. Jean
Vigo, 1931). © Gaumont.

understanding too of the local infrastructural benefits that can be bought by international sporting prowess. In 2003, Sport England (as the council was renamed) published a set of guidelines for the correct management of discipline in swimming pools.[11] It describes procedures that should be followed in any circumstance. It also alludes to a field of abject responses to behaviors perhaps alien to Taris and his healthful engagingness. "Procedures should be established to cover faecal fouling incidents," the document reads, and "all staff should be aware of these procedures." After any such incident, the pool should be closed, as should any pool whose water treatment is linked to the fouled pool, and "bathers should leave the pool and shower thoroughly." The writing is impressive in the way it steers clear of a sense of nightmarish insult to civil decorum around personal cleanliness brought about by such spontaneous pollutions. "Diarrhoeal fouling," the handbook continues, "is likely to contain bacteria and viruses." This infectious area is, ostensibly at least, the real area of concern for the guidelines at this point, and time is taken to detail the risks; particularly of infection from the oocysts of protozoan parasites like cryptosporidium. Ultraviolet light and the sensible application of hypochlorates are suggested as the counter to any such calamitous breach of hygiene.

The accidental fouling of the tepid, oleaginous water of the municipal pool is also a social figure. It isn't difficult to imagine the visceral recoil by swimmers, or the sudden, shouty marshalling of swimmers by attendants in clearing the pool. The grumbling at inconvenience and the plosions of disgust and contempt made by swimmers are easily imagined too. Faced with dealing with social meanings, it is perhaps understandable that the language used by the authors of the Sport England guidelines stolidly refuses any hint of scatology. Anxieties about disapproving attachments of blame around such an intimately registering failure of civic continence are passed over. Dejection, anger, approbation, remorse, embarrassment, contrition, and accusation suddenly appear as unspoken details of a violation of public virtue.

—————

What this attention to fouling also announces is another interpretive poetics of criticism concerning the very status of civic water as an architectural medium. It starts to gesture not so much toward the subjective effects of the pool on its users, but rather toward the possible subjectivity of the pool itself. Agitated into its *vocalese* only by its sociable use, at other moments the pool lies handsomely still and reflective. It hums its own gently regular refrains of pumps, heating and filtration systems, air-conditioning units, minutely buzzing fluorescent lights, flushing cisterns, and the clunk and rattle of vending machines. At rest, it is a modern aural poem of health and order, in fact.[12] The temptation to lend latent sentience to the pool, either in its repose or in its noisy participation with social physiques, is almost unbearably maudlin. Yet, certain habits of architectural criticism work at exactly this threshold.

The view of an organic condition for architecture is a varied and familiar staple. For the central tropology of Frank Lloyd Wright's rhetoric of the development, purpose, and aesthetics of architectures and for the character of relationships in the society that produce them, images of organicity are literally vital. As promoted both in Wright's own writing, and in the celebrations of his work by historians like Bruno Zevi, the term *organic* is announced not just as a distinction from the architectural habit of grafting-on "pseudo-classical orders," but as a way of coinciding the eternal principles of Nature and the American people.[13] Wright's desires to establish what he saw as an intrinsic connection between a national domestic idiom and the daily democratic charms of Americans identifies land with people in an organic relationship at the poetic site of the building. He extended this view to the forms of the cities and communities that should emerge from this social contract.

However, writers have recognized other conditions for that metaphor of architectural organicity. Peter Collins, for instance, wrote in the 1960s of a long-standing set of biological and gastronomic analogues that have quietly sustained an unstated structure of modernity in architectural discourse.[14] More recently, Mark Wigley has exercised digestive conceits to attempt to grasp the social-philosophical dimension of architecture.[15]

4.12 Ben Johnson, *The Unattended Moment*, 1997. Acrylic on canvas. Image courtesy the artist.

Tropological approaches like those developed by Collins and Wigley have been able to marginalize the ideologically driven notions of an evolving, self-regulating, evolutionary capacity for architecture, and by extension, society.

Water may well have a memory, as Jacques Benveniste once famously announced to a medical audience.[16] Le Corbusier, speaking of submarines, saw in them the architectural image of "organisms … perfect entities"; architectural commentator Charles Jencks has, with others, turned to the poetics of genetic coding and selection as a metaphor for historical process and a self-evolving agency.[17] And the new interest in Gilbert Simondon's reorientation of philosophy toward an understanding of the agency of inanimates such as machines and buildings shows a new sophistication being brought to the notions and metaphors of organic self-authorship.[18]

The pool becomes interesting here. Its own sound might start to approximate that of recognizable speech of the kind suggested by Joyce and Vigo. More, the pool is constituted as a body of water by its legally insisted-upon requirements to embody a responsiveness to matters concerning health, and specifically infection. As a special form, then, the pool prompts a point of paradox, rather than a simple, oscillating indecision over the architecturally substantial meaning of water. Water stops being a nonarchitectural material that is simply contained and managed by architecture on the instant that a photographer or painter captures a glamorously convenient reflection in it or when the poetics of redundancy that are cued by a pool, empty of water, are animated.

In its murmuring self-regard, the pool's own digestive facility manifests a theoretically elusive capacity. The sound of its plant aurally allegorizes the need for filtration, for the circulation of chemicals, and its ability not only to destroy and dissolve unwanted organisms, but also to eat at itself, breaking down its own integumen and constantly demanding its constant renewal. In the vast literature on the necessity for a conscientiously regular care and management of the means by which domestic, commercial, and municipal pools sustain themselves, that has been produced in Europe and America since the mid-1930s, and especially during the 1970s, a visual language of symptoms has developed. Images of cracked and decayed concrete, compromised linings, flaking paint, corroded plumbing, exposed and rusting rebar have illustrated a set of expectable difficulties as well as options for remedial courses of action.

A further set of images has emerged in advertisements for a range of products made for medical support of the pool. Within this, an unsurprising repertoire of *décolleté* physiques, sexually available in their Virgilian settings, is supported by an entertaining practice of neologism that supplies a further dimension to the pool's own *vocalese*. Thixoclor, Velpruf and Cemcol paints, Vanodine, Alquimycin, Elbadyn, Baquacil and it may be necessary for a structural lining of sprayed reinforced concrete, 75 mm thick to be specified. However, such a lining on the walls will reduce the inside dimensions of the pool, and if on the floor will reduce the water depth.

The surface of the concrete should be prepared by high-velocity water jets as previously described.

Further information on the use of cover meters is given in Section 10.22.2.

There may be some evidence of debonding of either the rendering, screed or tiling while the work is still in progress; recommendations for testing the bond are given in Section 7.8, and reference should be made to Section 10.26 for remedial work to this type of defect.

REMEDIAL WORK TO EXISTING POOLS: TRACING LEAKS AND INVESTIGATIONS

10.20 Introduction

Remedial work to existing pools is usually initiated because leakage is found to be taking place and/or serious visible defects have appeared (Figures 10.6–10.7).

Figure 10.6 Defects in floor of old open-air pool. Courtesy, Colebrand Ltd.

4.13   Damage to the concrete lining of a pool.
       Image courtesy Colebrand Ltd.

Flexicote Cleankill biocides, Flexometal changing booths and footbaths, Brakol alkali-balancing agent—these are all part of a consonantal iconography of cleanliness, producing "not just filtered but polished water."[19] The language is extensive, with new words coined to describe specially retarding cement mixes and groutings, etch-resistant enamel surfaces, and, in the application of sprayed asbestos-concrete, possibilities for acoustic regulation. Over time, the proliferation of these proprietorial nouns has been further paralleled by developments in underwater photography. New photographic techniques used to study and refine competitive swimming strokes in the 1930s, and which led directly to differing national superiorities in successive Olympiads in the first half of the twentieth century, came to eventually allow for the sharper resolution of the human figure underwater. It is possible here to imagine the different arms of commercial and athletic cultures

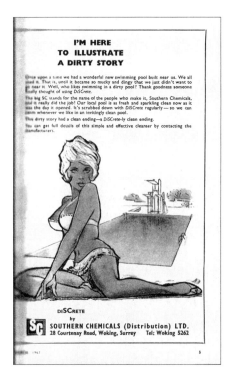

4.14 Advertisement for DiSCrete pool cleaning agent, ca. 1963.

4.15a Advertisement for Algimycin algicide.

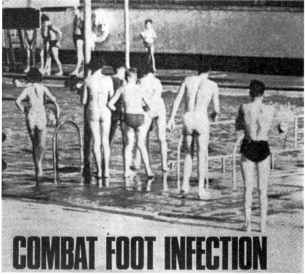

4.15b Advertisement for Evans Vanodine
pool products. Photographs taken at
the glorious but now sadly demolished
Ormston Baths, ca. 1960. Courtesy
Evans Vanodine.

colluding in a historically concerted effort to efface the *visibility* of water as medium while at the same time it was becoming more and more phonemically present.

When still, the pool provides a surface for a narcissistic identification with its surrounding buildings, and acoustically supplies a predictable and comforting regularity. In use, it seems that the irregular syncopated staccato of its sound is commercially simulated by a rhythmic glossomorphosis in the aggregate names of its health-sustaining medicines. This would only be appropriate. The swimming pool differs from any other type of architectural pool in that a worthily aggressive and desirable pollution, that is to say, its very *inorganic* chemicalized state is that which maintains its condition as an organic social forum.

4 Peckham Boy

It is worth remembering a foolish moment when, in reviewing the architecture of Italian health cultures in the 1930s, Philip Morton Shand, in celebratory fashion equated the priapism of a diving board designed by Gerhard Bosio in Florence, with the "commanding forefinger of Il Duce."[20] With his use of plans and models and indications of heli-cars, Frank Lloyd Wright was not alone in offering the lofty vantage on the broader scope of his architectural proposals as a motif of command. Le Corbusier, Steen Eiler Rasmussen, Leni Riefenstahl, Antoine de Saint-Exupéry, even the more socially empathetic Sidney Gilliat, in the opening sequence of his film *London Belongs to Me* (1948)—each has assisted in the elaboration of a scopic convention for the modernist architectural purview that is not based in any conception of participation.[21]

It is easy to imagine the chop of the blades of Wright's heli-taxis isolating a viewer from the sociable hum of his Broadacre City projects below. The sense of an architectural imagination cocooned from involvement, deafened to the irregularities of society by the roar of aircraft engines, seems essential to the figuring of objective detachment. For Vigo, however, the sound of water is a means of depicting architecture and the potential for a cheeky civic politics. The participatory involvement of the architect, rethought not as bird or flaneur, but rather as swimmer or diver, observed and observing, comes forward as the figure of a criticism. The former, the swimmer, announces a figure for an explorative participation in the substance of an architecture; with densely syncopated and extensive rhythm as the key to its perceptions. The latter, the diver, announces an elevation certainly, but one that remains connected to the sociable sound of the pool. The diver, however,

4.16 Owen Williams, Pioneer Health Centre,
      Peckham, London, 1935. View of the
      diving boards. Photograph © Wellcome
      Library, London.

embodies an ability to convene an expectant and instructive silence. These are two subject positions joined by flight, or indeed by falling.

———

In a rather shy essay for the *Architectural Review*, detailing the shifts and changes that had led to a rejuvenation of the swimming pool as an object of interest for architects internationally in the later 1960s, Bryn Jones thought it possible to identify two kinds of spectator at pools. He wrote of "the one who has come to watch an event, and the one who has merely dropped in to watch general swimming."[22] There is a kind of motivation in Jones's attentive observer of "swimming-in-general" that may be read as innocent or not, charming either way. At the time of Jones's writing in 1967, John Dryburgh's classically detailed pool at Cardiff was one of only three major pools that had been developed in Britain since the end of the war. Now demolished and replaced by a multiscreen cinema, this long-treasured amenity was sponsored by a municipal desire to stage the sporting spectacle of the 1958 Empire Games.

Dryburgh's pool, and indeed Bryn Jones's comments, arose at specific points in the cultural trajectory of shifting meanings found at the intersection of health and the nation's public baths. This polar shift was fully effected by the 1970s. There existed a kind of medical anthropology, one that from the 1930s had served as one of the great myths of healthcare, and which aimed to secure physiological health through observation, advice, and a faith in the prophylactic benefits of communal exercise. By the 1970s this had changed to a psychological sense of well-being conjured by a notion of spectacular leisure. Each of these, in its own way, refers to a discursive construal of the architectural purview.

Owen Williams's designs for the Pioneer Health Centre in Peckham, south London, with its Olympic-scale pool, built in 1935, are illuminating here. Respected now as a great landmark in experimental healthcare in the UK, and the focus of enormous popular interest at the time, it is interesting to see that in the earliest surveys of the emergence of modernist architecture in England Williams's building found itself slighted. J. M. Richards, at that time finding his place as one of the leading commentators on the ethical values of modern architecture for an English-reading audience, was keen on it. Noting small crudities and clumsinesses in what is otherwise a veritable eulogy, Richards celebrated the ways that all compartments of the building "merge into another, visually when not actually."[23] His approach, in keeping with much popular coverage at the time, was to

4.17

4.18 The Peckham family triad convened
(top) and distressed (bottom),
illustrating the ideological function of
the glass wall as seen from views
of the café looking onto the swimming
pool at the Pioneer Health Centre.
Photographs © RIBA Library (4.17),
© Wellcome Library, London (4.18).

return to the image of the building as a ship moored at night. Here, Richards evinces a ready critical familiarity with the Corbusian poetics of naval architecture as a principal motif in modernist approaches to building design, both in detail and in broader image.

The degree to which this metaphor was popularly mobilized by others at the time is remarkable. In describing "green water on one side of every room," and likening the experience of the building to that of being on a ship, an anonymous reviewer in the *Observer* newspaper exactly reflected the prevailing critical language turned to by many others.[24] Archival footage of the forms of recreation promoted at the Pioneer Health Centre certainly remind one of deck quoits and loungers. But, as Richards pointed out, the uninterrupted lines of sight, especially in the ways that they produce a sense of physical spatial continuity, also had a function for observation, if not exactly regulation. It is important to note that in the contemporary literature on the building, just as we find references to ships afloat in a verdant *terrain ideal*, we find references to the "glass cages of the entomolygist."[25]

Henry-Russell Hitchcock, on the other hand, seemed less convinced of the aesthetic case presented by Williams's building. Arguing for British leadership in modern architectural design, in his catalog essays for the 1937 exhibition *Modern Architecture in England* at the Museum of Modern Art in New York, Hitchcock found himself more impressed by the dazzling brilliance of Berthold Lubetkin in the way he established a perfect stage setting of the penguins' "music hall turns" for their enormous popular audience, in the pool designed for the London Zoo in 1933. Lubetkin was responsible for the design and execution of the pool and medical complex at Finsbury Health Centre in 1935. It is true that this building has been more enthusiastically received as architecture than Williams's contemporary building in Peckham.

The photograph of the Peckham Health Centre that Hitchcock included in the MoMA catalog makes the building look a squalid and paltry affair when compared with the refined accomplishment that was lent photographically to other buildings shown. In Hitchcock's defense, however, it should be said that the delinquent unruliness of the rear of Williams's buildings is something of a signature of his. The massively substantial loading bays at the rear of buildings he designed for Boots Chemists, for instance, articulates a propinquity between an unwatched and tatty ephemera of workerly bonhomie (cigarette ends, loosely stacked boxes, bits of string) and the shiningly grand disinterestedness of his fenestrated facades. Hitchcock's visual diagnosis of what he specifies as a formal failure on Williams's part seems based in a characteristic failure to understand the varied social ambitions of the modern buildings he claimed an interest in.

65 WILLIAMS, Sir E. Owen: Pioneer Health Centre, Peckham, London, 1935

4.19 Rear view of the Pioneer Health Centre as it appeared in Henry-Russell Hitchcock's catalog to the 1937 exhibition *Modern Architecture in England*.

4.20 Owen Williams, rear view of factory for Boots Chemists, Beeston, England, 1933. Photograph © RIBA Library.

Williams's building was one designed with a powerful grounding in the rhetoric of living in mind: living architecture, living society. It was very influential in this. During the period of the founding of the National Health Service in Britain after the war, advanced medical-architectural thinking made demands for the design and construction of "buildings to satisfy the needs of medical science."[26] This was stimulated dually, as David Goodfinch argued at the time, by the need for a sociological survey of prospective patients, and a recognition of the potential role of appropriately organized architectures in "making the public health-conscious."[27] Much of the material produced in promoting the requirements of the National Health Act (1946) cited the experiment at Peckham as a pretext. However, as has been argued since, the strictly curative aims of National Health provision, then and later, ran directly against the medical ethos of the Pioneer Centre.[28] The importance of this lies in a distinction between two approaches adopted. One, advocated later as the principle of the National Health Service, sought only to manage symptoms and diseases as they presented themselves at a moment of crisis. The other, the one envisaged by the staff at the Pioneer Health Centre, sought to develop a form of social symptomatology by facilitating both a medical and a lay awareness of an array of as yet imperceptible physiological problems, and to ground these in a discussion of the prevailing housing conditions of the immediate environment of this corner of south London. A preventative approach, based on this identification of as yet unrecognized maladies through the close monitoring of family groups, was detailed and widely publicized by the founders of the Pioneer Centre, doctors Innes Pearse and George Scott Williamson. Their methods and arguments were as contested as they were admired. The reasons for this resistance have been speculated upon: as an epiphenomenon of the sometimes demanding personality of George Scott Williamson, as a response to the demand that there be a complete separation of health services from medical services in any provision made by the state, or as a failure of the Peckham doctors themselves to provide any convincing evidence of the marked benefits claimed for the health and structural cohesiveness of the families who attended the center.[29]

Extant reminiscences about the short life of the Pioneer Health Centre and its impact on the families who organized much of their social lives around it during the years before and after the war leave little sense of an unfriendly or domineering regime. Those who used the place were fond of referring to themselves as guinea pigs or rats; and, in leg-pulling way, according to Williamson, they were quite unembarrassed to introduce themselves to him in this manner. Equally though, many who signed up to join the center soon left. The center's staff often gave as a reason for this a slight social terror at the prospect of the required annual "personal medical overhaul." Families were required to join the center as a group, in order to make use of its facilities and amenities. Individuals, as such, were simply not allowed membership. And, as a matter of regular procedure, families were medically examined together as a group. No treatment was offered at the center, only the identification of infirmities and the proffering of advice. The kinds of observation engaged in by the doctors emphasized a study of gait, carriage, constitutive type, general athletic fluency, and performance in recreation, especially in the pool. Moreover, and like similar American examples, the center was hoped to become financially "self supporting through its recreation."[30] A fee of a shilling a month per household was thought adequate to achieve this, in a catchment described by "easy walking distance." In this, the ethics of the center saw the well-being of the community more generally, as its subject.

Tim Benton, one of the earliest architectural, rather than medical, historians of the Pioneer Centre, has said that it is difficult to strike the right tone when speaking of the meaning of the place.[31] It would be easy to condemn it for the forms of surveillance it insisted upon, for the interpellative subjective effects of its architectural features, and how these were justified. It would be just as easy to simply equate it with other contemporary health cultures with more suspect political motivations, such as examples found in Italy and in Germany. In fact, neither architectural nor medical discourse has so far attempted to account for the other forms of association that the Pioneer Centre came to sponsor, if not overtly champion. The image of medics peering through glass walls at their swimming subjects, arms crossed, murmuring knowingly to each other, is entirely warrantable in the analysis of the Peckham Health Centre.

The institutional politics of the Pioneer Centre insisted on the importance of membership of a group. Moreover, becoming a member of a swimming club, a drama society, a discussion group, or even a whist drive brought with it responsibilities to that group. An ordinarily harsh and manipulative discipline is suggested in a note pinned up in the center at one point, which spoke on the meaning of courtesy, and the consequences that would be visited on anyone, and anyone associated with them, who showed any lack:

The group itself is responsible for the actions of all and each of its members. Any lack of courtesy or actual discourtesy, shewn by any members of the Group will draw the consequences on the whole of the Group. It is discourteous, for example, to forget to replace, in its proper place, the key you use. It is discourteous to use the sprays for any other purpose than that of cleansing the body before entering the baths; it is discourteous to the other members who are paying for the wasted water. It is discourteous to enter the pool with dirty knees or feet or hands or face. It is discourteous to take the bathing dress off except in the spray box, for then you wet the floor for those who follow you. It is discourteous to enter the pool before the bell ceases ringing, and it is equally discourteous to be in the water when the bell ceases ringing.[32]

The note goes on to reprimand about lateness, the wearing of incorrect attire, or any other failure of common manners. The delinquent episodes that prompted this stern lecture on civil decorum are unknown, the note itself anonymous. This form of interpellative arm-twisting was accompanied also by a fantastic image conjured of the passivity, the biddability of the community who used the facilities of the center. Innes Pearse noted her view of the "useful simplicity" in what she regarded as the study group:

The choice of our particular populace has been a happy one in that the members are characterised by a directness and simplicity that enables them as, the experiment proceeds, to pick up language to describe what is in their experiences, while remaining quite unresponsive to ideas outside their ambit. There is therefore no danger in this setting, as there would be among intellectuals, of their adopting ideas which have no counterpart either in their experience or their capacity to experience.[33]

This perceived, animal-infantile simplicity of the "uncomplaining man in the street" might appear to speak as much of cultural naivety as totalitarianism in the management of the Pioneer Centre. However, an underlying anxiety, recorded elsewhere in other instants (see figure 4.17), lends a meaningful counterpoint to the simple happiness performed by the subjects of Peckham, and the parameters of what Innes Pearse took to be "their capacity to experience." It is something that Michel Foucault might have recognized for his theorizations of discipline. Built around notions of goodwill and expertise, happiness signified at Peckham an image of pleasurable social democracy. As part of a compacted acting out of caste distinctions, the members of the center, who came from a range of class backgrounds, appeared to align themselves with given images of what it was possible to experience. The heroically tolerant figure of the man in the street is staged here with the insistent, disciplinarily rebellious expertise of Pearse and Scott Williamson written through it, turning people into what was *known*. Some of the center's users have since advanced a view of the Pioneer Centre as simply a pleasant and convenient social club with a pool and clinic attached. This view possibly indicates a kind of knowingness toward the work of this plenum of expertise in the formation of social democracy.

Yet, all this said, the Pioneer Centre was also organized, quite conspicuously, around considered notions of self-service and self-regulation. These were evinced especially around the use of the pool. "The kids who made most progress [in swimming] did so by watching the bigger boys. I always closely observed Ron Moody and George Goldstone": this is one representative memory of other forms of the less regularized kinds of instruction that became established at the center.[34] With a membership of three thousand, even at the quietest times six hundred people, a large proportion of whom were children, could be expected to be making use of the building. With such busyness and so very few staff, it is hard to expect the success of any too restrictive a regime. In fact, members of the medical staff were keen to report, as one of their most important findings, that *facilitization*—simply making the pool, bicycles, or roller skates available for use—fostered a desirably ad hoc and creative mode of self-instruction. If this also led to people finding corners of the grounds to light

4.21　Norman Howard, the very different
sociabilities imagined by Owen
Williams's earlier and grander pool
design at Wembley, London, in 1934.
Image © RIBA Library.

fires and cook sausages, to bump or graze themselves or to engage in some episodic gang-warfare or casually experimental sexual promiscuity, then this was taken as part of a necessary, and, perhaps more importantly, *documentable* facet of the sociability of learning. The visual regimes of Peckham, where "no activity is necessarily isolated and everything taking place is visible to the scientist and the members," were made available for the furtherance of a particular mode of medical epistemology, but they also allowed for, even sanctioned, less deferent behaviors.[35]

The pool, and particularly its diving boards, figured centrally in this licensing. While architectural discourses on the Pioneer Centre, photographically at least, have primarily concentrated on iconic still-life renderings of the broadly heroic front facade of the building, on its less imposing rear aspects and views of and from the array of diving boards about the pool, these are countered by the overbearing weight of emphasis, in the archival records of the center, on images of the noisy and unruly sociability of the place. More, such documented anecdotes foreground the ways in which it was struck through by all kinds of other acoustic organization. The geniality of card games, the squawk, scolding and patting of children, the rumbling of roller skates on the roof, communally heard radio broadcasts, applause for the center's dance band or a bit of amateur dramatics, the odd scuffle, the kinds of gossiping conviviality that passed on all sorts of information ranging from the value of inoculation to the romantic lives of the adolescents—these differing acoustic signatures not only constellated the center as a social space, they characterized the medico-anthropological *function* of the place.

Among these noisy features, however, it was the pool that was both seen and heard from everywhere in the building and which represented the formal cohesiveness of the building. From the restrooms, library, and cafeterias, and even the consultation rooms on the second floor, it was the pool toward which attention would

4.22
4.23
4.24 The Pioneer Health Centre offered a number of possibilities for noisily informal association in the café (4.22, 4.24) and grounds. Photographs © RIBA Library (4.22), © Wellcome Library, London (4.23, 4.24).

regularly and easily drift. Though clearly keen on the ideological functions of the center's glazing in producing the visual appearance of fluidity, George Scott Williamson also understood and celebrated its spatial workings in different ways. He offered the joyous architectural image of him frequently pattering down the flight of sixteen steps from his consulting room, and the sense of being "engulfed in the action which is going on."[36] The kinds of encounters produced by such fervent and euphoric dashes supplied for him the more telling indices of well-being in the health-focused community that he was so keen to foster. In his public and not-so-public comments, Williamson would speak of an excited father proudly reporting on the way that "Young Johnny" could now dive from the second board, or a mother telling of her delight in watching her entire family, husband included, playing in the pool together "as children."

The social reserve and deference that seem to have pervaded English society at all levels during the 1930s make it difficult now to read the body language of the period. Even for those with a cultivated familiarity with newsreel footage of the time, the intimate social physicality of Scott Williamson, when for instance interviewing a male club member for a promotional film, appears odd. Williamson's ease, and his interlocutor's obvious, physically coded physical apprehension, may be ascribed to relative levels of experience in managing bodies and people. At numerous points in the minutes and public, promotional statements and reports made by the

4.25a The family consultation was one of the institutional cornerstones of the Pioneer Health Centre. Photograph © Wellcome Library, London.

staff of the center, the physical unease of their subjects was taken to be as much a social as a physically organic malaise. At a time when vociferous complaints about the state of London's urban peripheries as an immediate result of wartime evacuations and the longer-term effects of a renewed dash to the much-vaunted healthfulness of the suburbs sponsored by the building of the underground transport system, the moment of the family consultation and its conversational tone at the Peckham Centre seemed able to equate remedial approaches to the integrity of the urban family with the protection of the stability of the population of the city itself.

The discursive and observational chain of relationships between pool, family consultation and community, took the principle of apparently increasing physical confidence as the key diagnostic of improved health. This was coupled, especially in Innes Pearse's contributions to the medical approaches of the Centre, with a concern for sexual well-being. Her technical specialism in sexual health focused less on transmittable disease than on the ethological development of a well-adjusted heterosexuality. For her, twenty years before the radical change

4.25b The pool offered the opportunity to informally observe the gait and carriage of the Centre's clients, while also allowing the occasion to watch "swimming in general." Photograph © Wellcome Library, London.

in the official vocabularies and concepts of healthy sexuality ushered by the Wolfenden Report and forty years before the video was made to promote Bronski Beat's single "Smalltown Boy," the pool offered an opportunity to study the ways that viable heterosexual sociabilities were managed and promoted in the community. For Pearse, it was the togetherness of a successful marriage and "being in love that changes you and makes you able to function more fully."[37] Big on the idea of "the manliness of adolescent chastity and womanliness of adolescent virginity," Pearse's remarks indicate that the visual focus of the pool could be one conceived around

4.26 Queen Mary and Innes Pearse
encounter a toddler during a visit in
1948. Prime Minister Clement Atlee
is in the background. Photograph
© Wellcome Library, London.

the successful imaging of masculinity at Peckham. "Experience teaches," she wrote, "that the healthy adult male's need is for that mutually elective female who can further the maturation of his own maleness."[38] From her work, more generally, it can be seen that what the pool and its noisiness represented more than anything was the opportunity to monitor visually and aurally, to study, to correct, and encourage, even precipitate the modes of adolescent male ardency that were thought appropriate to the future flourishing of "organic" society in Peckham.

The point to highlight in this account of the Health Centre, and it is one that applies to Greg Louganis's bearing as he left the pool in Seoul, is that the form of modern, modernist medical look (it isn't yet a *gaze*) that fell on these Peckham families was one that concerned itself with a *social* understanding of symptomatology.

5   Good Grief

So far we have a picture of the economy of the modern pool as it bears on moral questions, physiological, pharmacological, and perceptual questions, questions concerning the potential agency of the pool itself and subjective identifications with it. National cultural issues emerge here, issues concerning sex, sexuality, and the domestic arrangements thought proper to them. We have seen too the ways in which each of these notions is figured as a contribution to a dense and complex aural architecture that produces the pool as both a faceted *and* a discontinuous entity.

I'd like to turn now to an image of diagnostic sound. As a medical symptomatology and as a form of local American, domestic architectural history, the course of Frank Perry's 1968 film *The Swimmer* describes, in an arc, a dense political allegory of transition. This trajectory is one from subjective integrity to utter subjective disintegration. In some senses, Perry's film and the Pioneer Health Centre alight on the same territories in the ways they are able to bring forward unrecognized maladies for inspection and comment. Based on John Cheever's short story, set in the affluent Virgilian landscape of suburban Connecticut, Perry takes the private swimming pool and its alcohol-floated consolations as the centerpiece in a set of tableau variations. Burt Lancaster is Ned Merrill. As Merrill's hallucination of himself as a pillar, a very special human being, "noble and splendid," unravels episodically, it becomes clear that he is not well and that he doesn't know this. At the start of the film, to Marvin Hamlisch's saturated musical scoring, Merrill strides out from an American *sous-bois*, as if from nowhere-nature, barefoot and clad only in neat black trunks. Over time, and through the

4.27 The gait of Ned Merrill in his
      psycho-allegorical element. Still
      from *The Swimmer* (dir. Frank Perry,
      1968). © Sony Pictures.

discourses of and about his swimming, it is revealed that Merrill, in a delusion, has managed to forget that despite his practiced and polished charms he is a bully. Condescending, duplicitous, mendacious, vain, and adulterous, his children despise him, his family has left him, he has lost his home, he is in debt and bankrupt. He carries these facts, unknowingly, in his flawlessly athletic physique, as terrifying stigmata. He is met only with aggressive hostility, pity or, most revealingly, a nostalgic and superficial bonhomie—"You son of a gun." He understands none of it.

  As an image of America's place in the world, as a function of political unrest and deep dissatisfaction with the indignities of wealth and jaded bourgeois cultural mores in the late 1960s, where there "isn't even anyone to flirt with," Perry produces a dominating conceit of the suburban pastoral. Sipping gin in an epiphany, Merrill realizes a private suburban river, comprising pools and stretching across the county to his home. But the film also essays a probing account of relationships between health, hygiene, and class difference.

4.28 "Then up the hill and I'm home." Friends look on in concern as Merrill describes his possible route via the suburban swimming pools of Connecticut. The one pool he can't remember, but of course everyone else does, is that of Shirley Abbott, with whom he'd conducted a calamitous affair. Still from *The Swimmer* (dir. Frank Perry, 1968). © Sony Pictures.

Through images that embrace both the thirstiest Republicanism and the political eccentricities of nudist philanthropy, the complacent occasional civility of a particular stratum of longed-for suburban society is shown as poisoned by a simmering interpersonal violence. On perceiving this river through the bottom of a glass, Ned sets off alone to swim home, and unwittingly to face his denials. One thing that is striking about *The Swimmer* is its orchestration toward an overwhelming sense of sympathy for the plight of this tragically irredeemable figure. It remains a fact that even in the relentless end, in the most acute moment of his psychical calamity, and despite the thundering flooding rain in which he is caught, Ned Merrill is offered nothing that is baptismally cleansing. Nothing.

The irreproachable symmetries of Burt Lancaster's unquestioning deportment and competent musculature, as Ned, act a vehicle not only for the decorative distribution of towels and droplets (whether from pool water or flung cocktails), but also for the conjuring and confounding of typical moments of civil interaction. Lancaster's reputation as a physical actor, in appreciating and displaying the psychological nuances of the variously fraught visitations with Merrill's past lives, is challenged and sometimes defeated by the demands of the film.

The narrative of the film follows, with some approximation, the episodic structure of the *Aeneid*. And, while much needn't not be made of this—neither Cheever's story or Perry's film have much in common with Virgil's apple-polishing articulation of Roman civic politics as virtue unrewarded—it is clear that something of Numicius's sacred spring plays a part in the sociopolitical ambiguities offered.[39] As a figuration and undoing of the democratic common weal of water, a nothing at which to marvel, two crucial encounters staged around Burt Lancaster's physical and emotional architecture will bear some consideration. The gently pornographic staging of minor poolside villainies and inadequacies that represent the substantial bacchanalian target of the film are interleaved with telling scenes where water has no tangible, delineable form; this happens once at an empty pool and once at a public pool.

A concern with cleanliness is established at the outset of the film as its refrain. From Ned's reminiscences of the "transparent, light-green" river water that he'd swum in for hours as a boy with Stu Fosberg, and the filter that takes "99 point 99 point 99 percent of all solid matter" from the new and infantilized pool at the Hunsacker's, to the crushing second at the end of the film when the attendant at a public pool contemptuously demands "Spread your toes!," and Merrill, conforming and by then drained, limps with an abject grimace onto the antiseptic footbath, we see a metaphorics around which well-being is carried by feet. Not only Ned's sprained and dirty feet, acquired in a soteriological trek and which interpellatively draw the suspicions of the pool attendant's superiority, but also the manicured toes of Helen Westerhazy, which he playfully kisses, and those of his ex, Shirley Abbott, from which he dramatically extracts a splinter. This redundancy of feet, idle or damaged, feet bearing only allegorical weight, sharpens the case for a nonpedestrian architectural purview. It is the splash, not the tread, that is lent and maintains regular propulsive meaning. As the metaphor for architectural involvement, it is the animating function of the swimmer's actions, not the observant capacities of the architectural flaneur that count—and there is so much architecture to consider.

4.29 "Go on back and wash those feet."
Dejected, Merrill obeys the growling
pool attendant and spreads his toes.
Still from *The Swimmer* (dir. Frank
Perry, 1968). © Sony Pictures.

This splash is the discrete and clipped sound that describes the dreamed integrity of Merrill. Sharing the same structure of Vigo's skipper as he dived into the Seine, it is a sound with which, as be dives into pristine, glittering pools, empty of others, he is given to cathect, narcissistically. He, his name, and his physique are the pool, and it is the detailed convening sound of that splash of entry, and the aural regularities depending from it, in which he happily involves his self. This aural expectation seems to be established to further highlight those moments when it is neglected. After his sapping exchange with Shirley Abbott, his sexual vigor again rejected, it is possible to aurally read Lancaster's somatic portrayal of wounded exhaustion. We hear his ongoing state of decline. It is registered with strokes that are increasingly broken and irregular. When he unconvincingly reaches for the other end of her pool, it becomes obvious that the key register of his well-being in the film is this aural diagnosis. More, in recalling the moment she was reduced to tears in a glamorous Manhattan restaurant when he terminated their affair—"I also raised my voice. Did you really think you get rid of me to the sound of fingerbowls tinkling?"—water figures again acoustically as witness to the dreadfulness of his mental state.

However, it is another order of poolside regularity that, produced by the feet of a child, Kevin Gilmartin Junior, as he bounces on a diving board above a pool empty of water, that pulls Merrill up in a troubling glimpse of suicidal self-apprehension. Mistaken in his understanding that the boy was about to jump, Merrill rescues him from the diving board. An opportunity to speculate on the boyhood travails of the young Merrill is laid out. On his journey across the valley to his own home, Merrill encountered Kevin, a lonely entrepreneur selling lemonade at the gates to the Gilmartin estate, abandoned by his parents for the delights, respectively, of Europe and the manicurist. Ned's request for a swim, grudgingly conceded, prompts a paternalism in him; a paternalism that the film develops repeatedly as an image as of Merrill's brittle subjective cohesion. Seeing the Gilmartin pool as an unfilled abyss, Ned is moved to countenance the failure of his life-saving project, until it occurs to him that he could simply imagine the water. He walks the length of the pool with Kevin who imitates him in his demonstration of first crawl, then backstroke, then breaststroke. The sound of splashing water is replaced by Merrill's soliloquy on freedom. In a lecture on self-animating individualism, he argues the metaphoric superiority of individual sporting pursuits over the distracting social requirements of team games. Giving a political voice to this water, his palliative suggestion of the political virtues of an athletically romantic

4.30 "Well that does it. That really does it. My project is ruined." Merrill and the Gilmartin boy dream a pool filled with water. Still from *The Swimmer* (dir. Frank Perry, 1968). © Sony Pictures.

individualism to the disconsolate, lonely, and increasingly dangerous-sounding Kevin, darkens as Ned makes to leave. The tensions between Kevin's desperate exhortations that Ned stay and Ned's determination to press on, abandoning Kevin to his introspection, establish a stage where the epistolary structure of Merrill's most recent relationships with the world presents the empty pool at the Gilmartin House as an epitaph to Merrill. The pool becomes the moment of a displaced and reflective perception of his own end. It is in these moments of the denial of the appropriate function of the diving board and the supplanting of the absent sound of pool water with that of a formal discourse on American domestic value that Ned Merrill is precipitated by the aural apparatus of pool architectures as a discrete, perambulating, pathologically forgetful subject, utterly dislocated.

6 Plangency

"What's the matter, Mr. Merrill? Your friends' pools run out of water? How do you like our water, Mr. Merrill?" The second social encounter that breaks from the aural convention of the splash entrained by *The Swimmer* is a final and public one, at the Recreation Centre Pool. It marks an acceptance, of a kind, that Merrill will not be redeemed publicly for his perceived sins. It is the final leg of his swim home. He is filthy, tired and his gait is awkward, stumbling. Until now, the emotional-political ambience of his poolside episodes has been of a restrained malice, unblinking, precise, but regulated. Westerhazys', Grahams', Hammars', Lears', Bunkers', Biswangers', Hallorans', Gilmartins'; these territorial proper nouns, and Merrill's transgression of their nomic purviews, have so far figured in equations of the mentalities and sensibilities of identifiable suburban cadres with their moral relationships to wealth, propriety, and civic decorum. These names, these brands, and the regulations surrounding poolside etiquettes that they encapsulate, represent landmarks in the teleology of his disintegration and Merrill's great social catastrophe is reached across these thresholds of contained politesse. Struggling through the assaulting racket of the four lanes of traffic on Route 424, he eventually gets to the public pool, but he lacks the fifty cents for admission. He appeals to the charity of the ticket office in vain, and asks a suddenly recognized acquaintance in the queue for the cash. His benefactor turns out to be one of his creditors, a barman.

As he tries to enter the pool he meets with the undressed hostility of the pool attendant. "Can't you read? Go shower." Merrill finds himself there in a plebeian melee, visually and aurally confusing, jostling, rowdy, with no understanding of protocol. It is the last situation in which he wished to find himself: as Dr. Scott

4.31 Merrill's terror at the pool filled with others. Still from *The Swimmer* (dir. Frank Perry, 1968). © Sony Pictures.

Williamson at Peckham put it, "engulfed by the action that is going on." And, having endured the alien embarrassments prompted by the attendant's derisive inspection of his feet for the stigmata of fungal and other infections, he eventually makes it to the overwhelming sensual cacophony of the crowded water. Here water has no form, no capacity to reflect, no chance to hum with its own voice. There are no discernible regularities, only shrieks and disorder. Blinded, pressed, and in all ways worried by the garrulous crowd, as he reaches the end of the pool, gasping, he is met by a contemptuous committee of burghers. The comparison between the body of Merrill with the bodies of this board of yet more of his creditors is a powerful one. His physique, like his voice at this point, though evidently worn, is still elegant, serviceable, and detailed. They are corpulent, brutally muscular. So are their voices.

Merrill is barraged. Complaining of stinging eyes and stinking of chlorinated water, that very popular odor which describes the very possibility of this pool as a social forum, and distinguishes it from the simply filtered water of his erstwhile friend's private pools, he is held to account. "Some rich diet you got up there." "I used to send up French strawberry jam, American strawberries ain't good enough?" "D'ya hear that? He'll send us a check? It won't be worth the paper it's written on." "Those girls of yours were giving you the raspberry." "Had to be dee-john mustard." "We're decent people, trying to make a living." His physical, political, civic, and emotional accomplishment compromised completely in the face of this cadenced tirade on his failures of continence and responsibility, his health revealed in tatters, confused, Merrill lashes out.

Unlike Taris, Merrill's cathection with the public pool is not happy. The pool is not just a locus of social anxiety; he is ferally, nakedly fearful and, heading toward what he takes to be the safety of the nature from which he appeared at the start of the film, he attempts to scramble up an embankment.

The weather turns into a downpour. By the time he gets back to his home and the end of his journey he is an emotional and physical wreck. This is compounded by his realization that the house, boarded and locked, is no longer his, and the happy past is categorically lost. Given its time in the late 1960s, it is hard not to see *The Swimmer* as an allegorical account of a lack of self-awareness on the part of the American bourgeoisie on the international stage. Regardless, the film ends with a summative view of its argument that all forms of authentic bourgeois self-knowledge are crushing public agonies.

As a biography of disintegration and the role of the sound of water in providing a diagnostic account of that, it is also a radically politicized form of architectural history; one with a medico-theological soteriology at the heart of its poetics. As a polemical critique of the pedestrian, it gives us the swimmer and a calamitous view of the subjectivity of the architecturally constrained flaneur.

———

Marvin Hamlisch's score for *The Swimmer* might assist in superimposing the composure and fall of Burt Lancaster's Merrill on Greg Louganis in Seoul. But there is a more pertinent, more immediately available soundtrack. The previous Olympic summer competition was held in Los Angeles in 1984. This is where Louganis arrived as an international figure, securing several medals for himself and a reputation as a star US athlete. That summer was also marked by a something of a dance-floor sensation. Bronski Beat released "Smalltown Boy." Now a piece of iconic synth-pop, the song provides a famous, one might say anthemic account of violent homophobia and the urge to leave home. It is also about loneliness and the usefulness of crying. It was sung in Jimmy Somerville's arresting falsetto, and it is interesting to note how his haunting manner of singing made a new contribution to the very idea of the timbre and tone of an artistically socialist, male voice at that time.

The video that Bernard Rose made for the single is as well known as the song itself. Set partly in a swimming pool, Rose had Somerville play a spectator to "swimming in general." There is a seemingly autobiographical moment that registers a surge of excitement as Somerville's eager though misappraising gaze is returned by an apparently charmed handsome diver. Urged by his friends to follow up, Somerville seeks him out in the changing rooms, and encounters the boy who appeared so amenable to his interest in different civic company. He realizes his error in judgment and backs away. Later Somerville is chased and cornered by this boy and his gang, and he receives a beating. In its earnest way, this video represents a very direct encounter with metropolitan homophobia. At the same time, it seems to attempt to make itself ironically anachronistic in the way it describes the social character of the viewing galleries in the pool.

"Smalltown Boy" appeared as an element of in the popular structuring of an image of gay, male sociability that included films by Derek Jarman and Stephen Frears, a rapidly increasing number of plays, novels, themes in soap operas and, of course, music videos. Rose had already made the video for Frankie Goes to Hollywood's "Relax" (1983). The desire of that seems to have been to set about forging a public acceptance for a more private world of the social and sex life, the poetic and deportment of gay bars. "Smalltown Boy" is concerned with a municipal architectural realm. Rose's video shows Somerville as someone watching the swimming in a way that he shouldn't be, and he goes to explore the social and sexual possibilities he sees. So, the video describes a set of spatial-subjective relationships that, in the mid-1980s at least, still carried with them a connotation of prohibition and transgressiveness. The spatial politics of the video is to present that sense of absolute prohibition as an anachronism.

At a time when the popular aesthetics of gay politics were textured by an enthusiasm and optimism that seemed to be carried in the euphoric delight of Somerville's vigorous and bluesy falsetto, the regular narratives of homophobia were also collapsing into a miasma of terrors surrounding HIV. Conservative government campaigns in Britain at the time raised awareness about sexual health, especially about HIV after 1987, to an extent not equaled since, and in ways that went beyond Reaganite arguments for sexual abstinence. Rose's video and Bronski Beat's song contributed to this awareness, and to an opening up of the sexual inhabitation of pools and baths of all kinds to the acceptance of a general public. And this came about almost on the day that the HTLV-3 or LAX virus (only later named HIV) was identified and isolated.

The maternalization of HIV in the UK and elsewhere, which led to visions of the great matrons of BBC broadcasting casually sliding condoms onto carrots at prime time as part of a sexual safety campaign, is now part of the comic nostalgia for the period. Drawing that particular collection of personae into the debate managed effectively to promote a sense of responsibility in relation to HIV, its transmission and effects across society. "Smalltown Boy," in its moment, was a film that was able to bring forward elements of the social condition of a malady for contemplation.

———

In 1988 in Seoul, when Louganis added another resonance to the acoustics of the pool, he and possibly he alone in the world at that moment worried about his blood in the pool. He knew he was HIV positive, but only a few others did. With the support of his coach, Ron O'Brien, he returned to the pool to overcome this blow to his physical confidence and win the gold in the three-meter competition, as well two further gold medals. It speaks of a competitive fortitude. Red-badged with something like military honors, including the Robert J. Kane Award for American Olympic excellence, Louganis took the place afforded to such dominating athletes as a civic model and icon of Olympic social ideals. He was seen in the company of Reagan and Bush. When he came out, in a filmed public address to the Gay Games in New York in 1994, he only seemed to add to his prestige.

However, when in a television interview with doyenne Barbara Walters on *20/20* in 1995 he told an audience that he had been diagnosed HIV-positive six months before he opened his head in Seoul, the response in the United States was less enthusiastically accepting. The outrage at this particular failure of continence, as an act of perceivably irresponsible pollution of the sacred pool of competition, stands as a kind of allegory of civic freedoms in American culture, similar to Frank Perry's. In 1995, anger was not as vivid as it might have been. "Pure selfishness" by a man incapable of sacrificing "his own glory for the safety of others," as it was put by one, evinced a desire for contrition and repentance, certainly. But, the Olympic Committee's refusal to insist on HIV tests for participating competitors, and the Clinton administration's decision to waive such examination for athletes entering the country, may have helped in this, as may have Louganis's easy and graciously contrite manner.

The immediate response in some quarters, though, was to detail a fear of what blood in the pool actually meant. Was it a contaminant? Did he, in fact, bleed? In the days following his revelation, many columns were dedicated to the discussion of the dangers of HIV in the water and the modes of its transmission. Some were more educated and aware than others. Looking back, Louganis described a desperate attempt to contain the

4.32 Greg Louganis composes himself for
his medal-winning dive at the Seoul
Olympic Games.

breach in his physical integrity resulting from the blow, hoping that he could "hold the blood in."[40] The experience was as alien and surprising to him as to his audience. He'd occasionally clipped a toe on a board as he flashed past, that's all. At a threshold of unknowing, in bringing his hands together before his dive, convening his relationship with the pool in repose and the silence necessary to complete his action, Louganis couldn't know what was going to happen next. As an architectural figure, one whose competence was to be so violently brought into question, Louganis articulated tensions. Quite apart from the rigors of competition, he was also living with the recently found burden of HIV, the sense of deserving what he called his "faggot's punishment," and the physical toll exacted by a sustaining prophylactic cocktail of Videx, AZT, Tylenol, Percocet, Flagyl, Bactrim, and Demerol.[41]

At this threshold of transition from integrity to disintegrity, Louganis was at the edge of becoming an iconological representative of the cultural politics of the period. The discussions of his dive in 1995 focused much on the risks unwittingly run by the US team physician, James Puffer, who repaired Louganis's scalp. The majority of concern has been on Louganis's handling of this specific instant. In this, Louganis was not offered a form of redemption. He was *offered* understanding, and he *took* responsibility.

Writing in the *New York Times*, Frank Rich argued that this circumstance may be seen as the "rewriting of the history of AIDS."[42] The complexities of the ethics involved here, that is to say of what can and cannot be undone around Louganis's wound, haven't left him in the collapsed condition of Ned Merrill. Maybe, in part, that is because of Jimmy Somerville's place in a historical change in attitude toward homosexuality in the mid-1980s. There is only a sense of a character deserving of compassion. And, as with the subjects of the Pioneer Health Centre, the expert goodwill of the American press, the American Olympic organizations, and the American talk show allowed Louganis to publicly *be*. If it points to anything, it points to the polyvocal character of the pool, and its capacity to bear not weight but fluid discourse.

The simple fact that the HIV virus is thoroughly denatured by the pool is the only thing to have later emerged concretely and self-evidently from the sound of the water in Seoul. Louganis, in offering an architecture of integrity in his behavior, offered this as something that Frank Lloyd Wright's pursed-lip essay on simplicity cannot possibly contemplate.

5.1 Detail of Étienne-Louis Boullée,
proposal for a Bibliothèque Nationale,
1785. © Bibliothèque Nationale
de France.

5 Blush

What dreadful confusion, what dire calamities, when panic takes

hold of people because they apprehend some catastrophe.

—Etienne-Louis Boullée[1]

Embarrassment produces a fantastic nebulosity in perception. South of the river in London, we find a second-hand record store, and examine the role of embarrassment and anxiety at work in the custodianship of aural archives and vernacular democracy.

1   Naughty Boys and Archives

I want to get around to talking about a second-hand music shop on Lower Marsh in London. But to get at what kind of urban architecture the shop represents, I need to break it free from the obviousness of its position on High Street. I was once embarrassed there. Not terribly so, you understand; otherwise I might not be able to find the words to describe what happened. That is a thing about embarrassment: as a condition of radical uncertainty it often has an effect of unhitching experience from language. I'd say that in its relationship to esoteric knowledge, this record shop was conditioned by embarrassment, not just for me, but for everyone who ever went into it. In that sense, what I want to get at is the aural-architectural value of hot skin.

———

Buildings blush. We know this. So varied are the determining factors in the curing of structural concrete, for instance, in terms of its eventual lightness of color or the kinds of mineral deposits that might flush unpredictably to its surface, that some architects develop a kind of professionally self-preserving resignation in order to forestall anxieties about the future look of their buildings. A case in point is the way that the leaching of salts

5.2 Elizabeth Diller and Riccardo Scofidio,
      Blur Building. Lake Neuchâtel, Switzer-
      land. Photographs courtesy the artists.

from the poured constructional components around London's Hayward Gallery has precipitated a voluble, lay dispute over concrete's habit of self-expressive incontinence.

Another architectural scheme broaches similar issues. The management of architectural embarrassment is one of the subjects of the Blur Building designed by Elizabeth Diller and Ricardo Scofidio. A fog, sited on Lake Neuchâtel and produced in the context of the Swiss National Exposition in 2001, the Blur Building relies for its most powerful architectural effect on tens of thousands of tiny, hissing jets of water. They create a vast and enveloping mist. Complex barometry is involved in the way the building discloses itself: not merely meteorologically, but emotionally. The building's nebulousness produces an unsettling image of architectural knowledge. This is not only in terms of a revision of what constitutes the proper structural elements of architecture, but also in terms of the social uncertainties given to visitors to the building. The Blur Building also employs "braincoats." Aside from protecting the attire of its users from the watery substance of the architecture, these coats are electronically networked, able to sense and respond to one and other's proximity. They glow and vibrate suggestively whenever they register affinities or antipathies between individual wearers. They arrive at this by the comparison of cataloged personal information held on an electronic database. Red means we should get along well; green: no chance! In 2001, the way in which this artificial skin flushes, perfidiously betraying its wearer, allowing no control over the sudden social awkwardness it might generate, was entertaining. Today, given the ways that modes of association are formed and interrupted, quite literally colored, by a disinterested, archival, decision-making technology, it is chilling. Either way, Diller and Scofidio's fetishistic IT joke allows a point of access to a discussion of provocative spaces produced by the abrupt publicization of privately held views—views so private that they may be as yet unknown even to those who hold them.

The fetishism of Diller and Scofidio's architecture can be seen in the way it hangs on the excitements provoked by technological innovation. This kind of collector's interest in the affective sphere of sexy new machinery is something that can hardly be held against architects who are overtly concerned with materially articulating the potential sensibilities of modern space. Their fetishism may also be seen as the structure by which some occasionally unsettling erotic suggestiveness is precipitated. In this coy manner, comparisons between bodies of electronically archived information translate into unwarranted gestures in a social sphere, manifesting different kinds of interest and intensity as they do. This approximates to a kind of displacement activity.

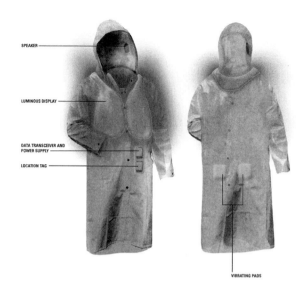

SPEAKER

LUMINOUS DISPLAY

DATA TRANSCEIVER AND
POWER SUPPLY

LOCATION TAG

VIBRATING PADS

5.3   Elizabeth Diller and Riccardo Scofidio, Braincoats. Waterproof jackets with luminous panels and vibrating pads, networked to an electronic database of information about the users of the Blur Building. Image courtesy the artists.

Sigmund Freud was interested in the socially and architecturally unsettling effects of hidden knowledge. In one way he considered these effects in terms of a kind of glow, too. His clearest comments on the phenomena of the fetish are organized around images of illumination, the "glanz auf der nase" (glow on the nose) described in a particular case study.[2] While defining the structure and psycho-history of the fetish, Freud was careful to indicate that, archived, tucked away in the willful mind of his naughty client in this case, was a particular knowledge, a knowledge that "the woman *has* got a penis, in spite of everything; but this penis is no longer the same as it was before." It was the hidden condition of this thought that allowed Freud's client to be able to project a shine onto the features of his desired one, in an act of will.

The glowing of the subjects of the Blur Building, on the other hand, occurs at the cruel and mechanical whim of the architectural workings of a database. The opportunities for social discomfort—whether sadistic, masochistic, or both—are manifold, and represent the comportment of knowledge in Diller and Scofodio's building. A question of the location of the agency of the subject in these processes marks a distinction between Freud and the Blur Building in attempting to understand the control of archival space.

———

Freud's understanding of the form and location of the internal objects of a private knowledge, as the dangerous source of potential embarrassment, is reflected in other psychoanalytical apprehensions of the emergence of childhood knowledge. In a remarkable essay, Melanie Klein psychoanalytically assessed a journalistic recounting of the libretto of Maurice Ravel's short opera-ballet *L'enfant et les sortilèges*. She connected it with a second account, one that related to the careful aesthetic organization of a particular house. This house, on the furnishing and decoration of which a great deal of thought and effort had been expended, was frequently left deserted by its artistic and much-traveled owner.[3] She paid no attention to Ravel's score, or to the character of available early interpretations of it by Albert Wolff and Victor de Sabata. Yet, Klein drew from the surreal literary narrative of *L'enfant et les sortilèges* a diagnosis of oedipal anxiety and a technique for the revelation of its primal scene.[4] Here, a boy, bored to disfigurement by homework, loses his already violent temper with the contents of his room (including, Klein notes, many part-object representations of his mother). He is then himself attacked and chastised, though eventually paid homage to, both by his furniture, which becomes suddenly, menacingly, and confusingly animated, and by the animals of the garden into which he'd scurried for respite. What impresses Klein in the libretto is the boy's understanding of the intimidatingly *giant* character of the objects of his room, their sublime psycho-historical scale, and what this tells the boy of his own ability to control and order a hostile world. This represents the locus of both his eventual ability to "conquer his sadism by means of pity and sympathy" and his ability to be recognized and redeemed by this act of gentle dominion.

Klein argues that what is central here is that, reminded of another fantasy—that of anxious oedipal loss and the dreaded castration punishment that necessarily attach to it—the boy summons forth for himself and accepts, though in desolation, the lesser abuses of the furniture: the phallic clock, the steaming kettle, the torn wallpaper, the slashed cretonne, and, notably, the shelved books and the armchairs. The worst of these violences, it seems, are not the physical blows he receives, but the upsetting of the order of knowledge that is effected by the little old man who emerges from under a thrown book. It is this, the unusually nihilistic and sadistic spirit of arithmetic, that scares the boy so: "Three times nine is twice six." In this image, the order signified by the catalog of the books is ruined.

This resignation of the child to a self-engineered punishment, precisely in the avoidance of the dreamed terrors of the privately held knowledge of an analogous future punishment (snip-snip), is a familiar Kleinian device. In her essay on infantile anxiety from which this assessment of Collete's libretto comes, Klein connects Colette's story to a further account—one by Karin Michaelis—which is equally staged in musical and architectural terms. The owner of the much-decorated house mentioned earlier had a pronounced appetite for fine art, but, it seems, no known brilliance of her own. This lack appears as the figure of a profounder sense of emptiness, for which her connoisseurial habit of collecting fine objects may have provided a wished-for, if rudimentary, armature: an infrastructural means of fulfillment. A crisis is precipitated for her by an irreplaceable painting lent to her by her brother-in-law, a nationally successful artist. When unexpectedly reclaimed by him, it leaves a gap on her well-considered walls, which seems somehow to significantly equate with her own occasionally constitutive sense of void. Depressed by this suddenly registered absence and perhaps by the view that her collection stood in for her own completed self-hood, she alights on the notion of attempting to paint directly onto the wall a substitute for the one lost. She orders up any amount of oils, brushes, charcoals, palettes, and other paraphernalia of painting. She produces there, to her own great surprise, a considerably impressive work. Disbelieving her own effort, she calls on the brother-in-law, the true source of judgment in such matters. Moved enormously by the work, the brother-in-law will not, however, be convinced by her that she had painted it: "If *you* painted that, *I* will go and conduct a Beethoven Symphony in the Chapel Royal tomorrow, though I don't know a note of music!"

Klein's impression of a sadistically organized, female equivalent of castration anxiety is all about the condition and psychical purpose of collecting. A sense of loss is articulated as a desire for reparation, and this is the replacement of the lost and irreplaceable painting—by her own efforts. What is also important, and less highlighted by Klein, is the incredulity of the brother-in-law. Seeing no guarantees in terms of evinced hidden knowledges or even technical predisposition, he preferred to think that another, proven, more skilled and experienced individual had produced the work. His critical paranoia is matched by the newly uncovered, as yet unrecognized artist's desire to prove that "the divine sensation, the unspeakable sense of happiness she had felt" in completing the first painting could be repeated; that the painting had a significance guaranteed by proven accomplishment or knowledge. As far as Klein's take on the narrative of Michaelis's story is concerned, the woman, Ruth Kjar, spent the rest of a skilled and productive career providing convincing evidence of a knowledge and capacity that could substantiate this questioned first painting.

These observations between Diller and Scofidio and Klein establish issues of disappointment, fear, and threatened punishments around the satisfactory demonstration of a comprehensive body of knowledge required to sustain and guarantee the archival significance of extant objects. What I hope they will help with is the suggestion of a critical value for the subjective effects of *overhearing*, in an overdetermining sort of way, and in a way that preserves for the meaning of the term *critical* something of the significance of the idea of a *crisis*. So, we shall explore something to do with the kinds of calamities that may be involved in becoming the subject of overhearing, and the kind of spaces, observations, and urbanities that this may produce.

2 Rare and Rotten

Gramex is a terrifically cheap and really rather learned second-hand record shop, not far from where I live in south London. It has been in existence since the early 1970s. It deals in classical repertoire and nothing else. It will sell you a laser disk, an eight-track cartridge, a cylinder, vinyl, shellac, a DVD, a video or CD, anything—but only classical. For London, certainly, and possibly internationally, it is unique. I have spent a lot of time there, though as I write, it is changing; becoming more recognizable, generically, as a commercial concern, and networking itself into the compacted trusts of online sales. Means of cataloging, buying and selling, finding clients—all these things are now in a state of radical review.

However, it remains a remarkable place for many reasons. In part, this is because of the clientele it attracts and the way they occupy the shop's very many, battered chesterfield armchairs. The place has a fusty aesthetic charm, falling somehow between a parody of a gentleman's club, an antiquarian reading room, and someone's parlor. This is only enhanced by the apparent insignificance of the architecture of the building itself. It seems barely to have an exterior. One would be hard pressed to recognize immediately that its walls and roof and basement were built in the later years of the eighteenth century. Its aluminum security shutter supplies a cherished anonymity. Just where the architecture of this building actually resides is something I shall here address.

The place is also remarkable for the fact that, despite what one may think on entering, and whatever it is that is remembered of a visit, no music is actually played there, ever. Part of the much-reprised social history of the place includes the information that there was once a plebiscite, a long time ago, and the clientele voted

5.4   Gramex. Author's collection.

no to music. But, despite this quiet, one might say that musically, the place is in fact, cacophonous, and deliciously discordant. This is a very striking metonymic effect; an internal, musical audition may be prompted here just by the sight of walls lined with disarrayed vinyl, and tables heaped with CDs and all interspersed with people, collectors, talking, or thinking, or reading—or, indeed, sleeping. For the fact is, in the more grossly material aural world, the shop sounds only of the slumbrous murmuring of discussions about music … and sex, and politics, and religion.

You may guess that I am describing a place that is, especially during the *longueurs* of a weekday afternoon, a place of retired men. The odors and tastes, the textures and the paces of the place rarely remain constant, but they are always, at this time of day, those of older men, and there is always someone there ready to say something (whether loving or charmless) about Victoria de Los Angeles, or the current Home Secretary or Archbishop of Canterbury. What surprises, always, are the ways and extents to which these rather dusty-sounding characters confound expectations of what they should amount to. This is a place that is interesting for its civics, its ideas of the custodianship of and participation in a musical culture, and its own stated attitudes toward how it sells, to whom it sells, exactly what it is that it is selling, the kinds of transactions this may allow, and how all this has been wrested from the hearty industrial capitalism of the recording industry. It is a curiously democratic place, its learning various, and worn variously.

Despite the blandness of its mainly uninterrupted glazed frontage, in fact in part because of that, behind that shutter what the building looks like more than anything else is Étienne-Louis Boullée's design for a national library. Published as a cultural-political proposal in 1785, it was not built. In his image of access to collected knowledge, there are questions about the representation of archeology and civic continuity, and how the ordering of knowledge came to found the rhetorical, architectural language of secular bourgeois democracy. It seems to prefer certain types of association, images of the state and its sublime relationships to its subject, its citizens. Look at the way the volumes, treatises, and dictionaries themselves seem formally to converse with the regular fluting of colonnades. And there appears to be an aural impression, perhaps a murmurousness, a type of representation of the sound of discourse in general, as it were, a *klang*, a kind of materially aggregated sonority of discourse, as well as the rustle of paper and fabric and feet. Fragments of this sound may appear to be occasionally foregrounded in abrupt perspective, other parts blurred, counterpointed, resounding and sympathizing with the scale of the building.

The relation of Boullée's architectural image of state to our shop also raises issues about the erotics of the intellectuality of old men, the modernity of their preferred modes of intimacy and association, intellectual and otherwise, and the particular structures of their need to know, with all the potential subjective crises and anxieties that are summoned and forestalled and inspected by them. It hints at a place in society, and its cultures; because what Boullée's image highlights too, and more than anything else, is the fact that, reveling in peripherality, the shop struggles vigorously to maintain its status as almost entirely unrecognized by broader social and commercial consensus.

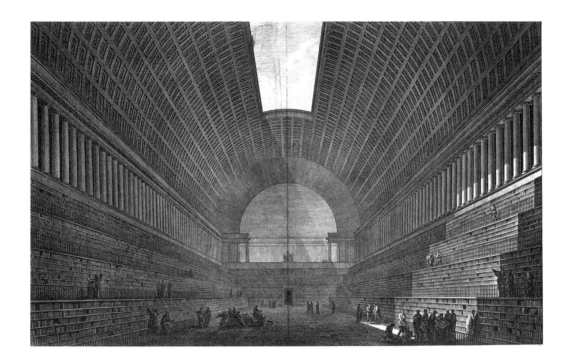

5.5   Étienne-Louis Boullée, proposal for
a Bibliothèque Nationale, 1785. © Bibliothèque Nationale de France.

Now, I want to put all of this somehow in a context of the culture of the early 1970s. This may suggest, as I
have hinted, a reference to theories of interpellation offered by Louis Althusser, and his presentation at that
time of the self-questioning response "Do you mean me?" to interpellative forces. But there is a rather more
pressing issue of musical autobiography. During the first few years of the 1970s, I really had no musical scene,
as it were, to identify with, to be plausibly interpellated by. I was caught between, on one hand, the popular

romantic piano repertoire preferred by my parents—Grieg, Chopin, Tchaikovsky—and, on the other, by the record collections of the older siblings of school friends—Be-Bop Deluxe, Yes, Nazareth, Emerson, Lake & Palmer. These are quite different worlds. It is possible to imagine an eleven- or twelve-year-old boy in 1973 engaged in friendship-risking and milieu-founding arguments about the respective technical facilities of guitarists like Jimmy Page or Steve Howe—nothing too involved, just a question of who wins. But, it is not possible to imagine me or my parents squabbling over whether it was Heinrich Neumann or Tamas Vasary or Noel Mewton-Wood who was most skilled at handling the Chopin piano concertos.

It wasn't until I turned fifteen in 1977 that I woke up to something that unmistakably called itself my own, which presented as something that was impossible for me to misunderstand—the Clash, John Peel, and the warm moment of self-recognition when in 1977 Peel's annual Festive Fifty stopped being topped and dominated by Led Zeppelin and Pink Floyd, who were swept away in a siren chorus provided by the Ramones, the Buzzcocks, and Stiff Little Fingers. That image, there—an adolescent caught culturally unaware in the transition from album-oriented rock to punk—is an extremely familiar cultural historical trope, as colored by redemption as anything suggested by Melanie Klein. In fact, as an image, it is a cliché: repetitive and perhaps not as performatively iterative as a person might like. It is something that someone of my age, gender, class, and nationality might be expected to say. As autobiographical material, it is scarcely unique, almost negligible. More importantly, its typical ubiquity casts a light veil over the musical and sentimental history of that period, allowing only certain things through, in certain ways and on certain terms—terms that are usually very faintly and uncomfortably comedic.

Much is left out of this description. For instance, although I listened over and over and over again to particular records in 1973, and must have given every appearance of being something of a fan of them, many of them I never actually found myself able to appreciate. I never really *got* Yes. It was like I couldn't hear something, something that I took to be more than yet somehow embodied in the music. Even the later music I was supposed to enjoy had this characteristic: I was never terribly sure about the Clash, preferring, I think, the idiosyncrasies of Alternative TV, Wire, and the Fall. This unhearable something is difficult enough to precipitate and identify, let alone lend significance to or account for. It is an excitement, or an intensity, or an emphasis—one of the qualia caught up in the success of a modern cultural form.

Those interested retrospections on that period of culture that have dominated since the very end of the 1980s have developed a special kind of attitude toward much music of the early '70s. It is an attitude that may be characterized as an acceptable, even desirable form of embarrassment: a type of embarrassment released from one of its defining features—that is to say, a real, self-conscious, writhing emotional agony. The paradoxes of this type of emotional denaturing, in terms of allowing for the consumable inhabitation of particularized forms of frisson, no matter how syrupy, are intriguing. And, in part, this is because there are so many varieties of relationship between sound, space, and embarrassment.

We should return to politics and religion. It may be said that Dame Shirley Porter, as someone personally surcharged with a £37 million bill for corruption and gerrymandering as Leader of Westminster City Council in the 1980s, was a person with some very real reasons to feel embarrassed. I recently watched some older footage of her. She was walking along in front of the Houses of Parliament toward a group of journalists, around the time of the breaking of these scandals. She had two friends with her, one on either side, and, as she neared, both she and they, seemingly intuitively and spontaneously, set up a barrage of babbling, made up of various bonhomies and cooeys and pointless jokes, remarks about the weather, this sort of thing. There was clearly strategy here. By laying claim to an acoustic territory in this way, they made it difficult for probing questions to reasonably demand to be responded to; speaking was already taking place. What they achieved there was a manipulation of an acoustic terrain, a wall if you like, and certainly an acoustically spatial constellation of political and legal and journalistic protocols that protected Porter's public persona at that moment from the embarrassments of ad hoc interrogation on the street. This broadly winking babble of greetings was a piece of urban political theater. It was something that was struck up, it seemed, solely to serve the purpose of establishing aural obstructions to inquiry, hindering access to knowledge and opinion. Principally, their action subvertingly exercised a principle of civility: "Excuse me, dearie, I am speaking."

It seems unlikely that Shirley Porter was, in fact, at that moment in any way embarrassed by the legal situation she found herself in; she had more than enough civic plaudits to keep her feeling culturally secure. However, then something happened. As she walked, she made a much more fundamental error of political deportment: she tripped. With that, the whole vignette altered its character. Not quite slapstick, but now she really was embarrassed. For fear of ending up on her arse on national television, as she stumbled and rather inelegantly recovered herself, momentarily revealing another fear and desperate self-interest in the process, she and her companions augmented the intense volubility of the babble that they had already established.

She blushed, you could see—perhaps—but you could definitely hear. Their racket started to take on the habit of automatic writing. In the face of potential subjective disaster, phrases and remarks became increasingly unhinged and incoherent, revealing and contributing to the crisis that she found herself figuratively plunging further into. The situation was compelling because it seemed that in an instant her own physical influency and incaution had presented her with an allegorical image of complete calamity to her dignity on many levels, and this translated into uncontrolled yet still successive speech. Almost aphasic despite its successiveness, her language became materially defensive: articulating only her occupation of space, and nothing further. In it, there was a shift from the instantaneity of complex, emotional crisis to a palliative attempt to restructure the anxieties contained by that image in spoken words, over time.

Producing a form of redemptive and material temporality, this successive babbling remained of more or less similar timbre, but it signified in completely different registers. One register aurally marshaled a space and exhibited a politicized form of control of it. A second register depicted only a flailing lack of control of the circumstances, which reflected critically on the order of the earlier moment. With a little imaginative empathy, it was possible to glimpse some interiority there, and a significant relation of it to some aural and somatic exteriority. Questions of signifying conduct, then, in the face of embarrassment and the pressures of the need to know, seem to be important at such times. Porter's misfortune in the precincts of the Houses of Parliament is typical of an embarrassment, and conventional in its form. Personal agony and somatic discomfort are obvious components of it.

3 This Heat

Other, rather older models are available for understanding responses to conjunctions of sound and embarrassment, which are different in their figuration of emotional content. In his *Life of St. Francis*, in a chapter titled "On His Ardency," St. Bonaventure wrote this of his subject:

Of the Ardent love that glowed in Francis, the friend of the Bridegroom, who can avail to tell? He seemed utterly consumed, like unto a coal that is set on fire, by the flame of the love divine. For at the mere mention of the love of the Lord, he was aroused moved and enkindled, as though the inner chords of his heart vibrated under the bow of the voice from without.[5]

That "voice from without" is an important thing here, but there is much more to this short passage. We have by now become familiar with the kinds of inner auralities suggested by Christian mystical experience; of messages that have been, as St. Ambrose put it, "heard without utterance, and without the sound of words." There is too Thomas Alexander Browne's remark on Bach's 48 Preludes and Fugues, which he said seem to "fit the harmony that intellectually sounds in God's own ears," and this may be allied with St. Theresa of Avila's observation on such resonant interior auditions, whether they be as words, music, or other kinds of sounds without material equivalents in the wider world, that it is "impossible to misunderstand them and their significance, no matter what resistance is offered to them."

In his essay "Ideology and Ideological State Apparatuses," Louis Althusser went out of his way to say that he understood this as a model for the concept of interpellation he was to advance, especially with regard to the phenomenal fact that such hailings hardly ever "miss their man": the one called always recognizes that it is certainly they who are called, and not another. The clichés around Althusser's name may now have formed an impenetrable wall. So, St. Bonaventure's words are to be preferred over Althusser's for particular reasons to do with the Lacanian psychoanalytic expectancies attached to them by a particular formation of cultural study. In what Bonaventure said about Francis, there is an erotics; the imagining of some treasured intersubjectivity—not just the one that comes with the territory of Jesus understood as the Bridegroom, but more vividly in a desire to communicate with Francis through a piecing together of his biography and the auralities of his inner emotional life. As knowledge, it suffices for a kind of ownership and propriety in a relationship to Francis: the maintenance of a personal, knowing, and loving discourse with him.

We also find a strongly articulated image of a somatic condition here. The image of fire for the perception of divine love is one of the key tropes of early Christian writing, and it is one that continues, and which may find a mundane parallel in a blush, with all the disturbances of space, temporality, self-worth, identity, and symbolic meaning that is entailed by this. The burning skin, the abrupt encounter with an object of particular desire, the lack of control of circumstances, the focusing on something that by definition exceeds one's intellectual grasp at the time and the sense of something it may be better to keep mute about, or to babble interminably around—all of these things seem to be shared by Bonaventure's description of Francis's animation and more everyday senses and sensations of embarrassment. Of course, where Bonaventure's words differ from Dame Shirley Porter's is in her apparent lack of pain and self-consciousness.

Melanie Klein's remarks on her artist's painting introduces us to the characterization of embarrassment—here, a range of issues about knowledge: about provenance, judgment, attribution, types of connoisseurship, a familiarity with technical languages, resources, and archives, and an ability to work with allusion and connotation in discourse. A kind of effect, which we will come to again in a later chapter, is produced in the overcontraction of sentences or phrases or figures, and results in a particular kind of signifying condition, a bit like a crisis. Julia Kristeva has referred to this effect as his effect as the *epiphora*, and she lent it a special significance. It is a moment, but one that may include repetitively allusive, anaphoric elements. Also, it is a moment, or rather an instant, when, on first sight of a given figure, all possible meanings, suggestions, and allusions present themselves *as* possibilities, and in play. It is the anarchic moment, short-lived, perhaps atemporal, which comes just before one or another adjudication is made about the nature of the terrain that stretches between an utterance or form of words, and a preferred poetic security. It is the moment before an organizing guarantee is sought, before a decision is made about whether this regulated route between one thing and another is going to be metaphoric, elliptic, paralectical, or whatever. It could be described as terrifying, vertiginous, or delirious. It could also be described as simply rather pleasant, or a bit uncomfortable. Epiphora might be given democratically egalitarian political implications, or it might not. Its similarity in structure to a moment of embarrassment lies in the unsettling rapidity of the images and possibilities it provides in a moment before order and good conduct and the material temporality of successive speech arrives; and, as such, it is not only prefigural but prejudicial and preliminary. It is exactly as understood by the chaotic terrors of the child in the vignette supplied between Colette, Ravel, and Klein, and it is a somatically registering instant of knowledge itself.

Epiphoric embarrassment is, it must be said, not one of the great emotions of Western literature. In a sense, its condition defeats narrative coherence. The strand of farce in a tradition that stretches from William Congreve to Evelyn Waugh, Anthony Powell, and on through to Kelsey Grammer has rather bullied the finer points of embarrassment—regarding it as a disaster to be observed rather than a perceptual position to inhabit. For here, in these instances, embarrassment and its physical manifestation may be precluded by wit and erudition. The sudden nebulousness of a world precipitated by an embarrassment may be reordered through speech. One of the further reasons for the minority of embarrassment as a literary device is that, while equally affective, unlike love or hatred or anger or jealousy, embarrassment is not an abstracting, object-defining, or metaphysical mode. It doesn't produce terrains in which objects can be seen in their settled integrity, terrains across which subject–object relationships may be held knowledgeably in place.

4 Calm

Gramex has an existence as a commercial architecture. *Without Rhetoric*, Alison and Peter Smithson's set of personal retrospections and short-term urban predictions, was published in 1973. It has a kind of landmark status both for its grasp of the concept of the *found* raw architectural material, and its use of the term *architecture autre*, as well as for its understanding of the socially predictive power of architecture in its relations to advertising and other forms of "pop" theory. *Without Rhetoric* has advertisements, and desires for advertisements and their worlds, as "visual telegrams with a specially loaded message about possibilities for the immediate future." Their faith in this status for adverts put the Smithsons, they said, in a position for them "to give form to people's aspirations at the same moment as they discovered that they had them." And, because of the ever warring, factional, sometimes tragic, and often treacherous character of architectural discourse in Britain in the early to mid-1970s, the Smithsons turned to the pursuit and modernist elaboration of a notion of calm.

They write:

When all have machine-energy—cars, transistor radios, light—to throw about, then the time has come for the lyricism of control, for calm as an ideal: for bringing the Virgilian dream—the peace of the countryside enjoyed with the self-consciousness of the city dweller—into the notion of the city itself.[6]

They go on to describe the ordering and control of "urban energies." Citing first an enjoyment of the patrician conduct of Le Corbusier's environs at Chandigarh, they write:

This is why we enjoy the true town-room for walking only, like the Burlington or the Piccadilly Arcades: the urban salon ten feet wide, detailed to be seen from three feet away and not fifty: expensive, highly controlled, not aggressive—only the door signs are different from shop to shop—civilised, mind-releasing.[7]

Terrifying. This describes architectural interpellation in its refinements. Working at the level of the somatic production, rather than the mere, vulgar *signification* of intimacy, the Smithsons' view of commercial urbanity suggests a cued inhabitation of space shaped in the sometimes delightful, sometimes disabling frisson of the possibility for blunder and physical and subjective calamities of all kinds: speaking too loudly or inappropriately, pointing, sneering, or eulogizing, knocking things over, accidentally whistling, or singing or hurrying. It is an architectural suggestion of physical and emotional restriction, and the pleasures of it. It is a version of embarrassing architecture, risking affect and affectation at all times. More familiar theories of urban subjectivity, such as those ventured by David Harvey or Edward de Soja, have become comfortably resigned to the fact that the critical negotiation of such space is a Baudelairean art, one adept at dealing through learned discrimination with the silent, haunting, judgmentally knowledgeable gaze. But this is a theoretical body well adjusted to alienating concepts of commercial space, as functions of capitalism, where polished service, and the correct and knowledgeable distance are necessary to the dignity of shop staff and clientele, alike. It is not something that is well tempered toward kinds of commercial space conceived preliminarily as sociable, rather than just social—that is to say spaces, like Gramex, that foremost demand association as the preliminary to commercial exchange.

———

Because of this obvious sociability, it took me two years to finally summon forth the courage to walk into Gramex and its potential punishments. Coincidentally, the much-liked pianist Alfred Brendel has made a comment that reflects on what I came to find there. Brendel, with whom it is worth sympathizing as someone who, photographically at least, appears to survive at the threshold of making some appalling gaffe every minute of the day, remarked of his performance of unheard aspects of music:

When I saw myself on television for the first time, I became aware that I'd developed all kinds of gestures and grimaces which completely contradicted what I did, and what, musically, I wanted to do. I then had a mirror made, which I put beside the piano, not making me visible all the time, but always there, in the room; unconsciously one noticed things. It helped me to co-ordinate what I wanted to suggest with my movements with what really came out. There are many examples of pieces where this is necessary. Things like the end of Liszt's B minor Sonata … there is a crescendo on one chord which one has to convey bodily, with a gesture. It's the only possibility.[8]

Brendel's performance notes suggest a kind of proscenial character to the shop. They point to the appropriate-ness of a grammar of gestural remarks to a structured aural context, especially where there is some require-ment for the physical deportment of an interpretation of what is found, musically. In Gramex, at the shelves, the interpellations are complex, and there is a continual sense of derealization, a seeing of oneself. Desire and the intellectual labor of exercising imagination, discrimination, and difference, and lending significance to these, as each title, each record sleeve or CD box comes forward, are signified constantly by verbal "gestures and grimaces," small grunts and remarks of contempt and appreciation and surprise. An essential aspect of an immediately adopted, minor architectural and archival practice of making heaps of potential purchases, these grunts are set against and contribute to the shop's humming aural texture of conversation. Thus on stage, one occasionally also has to fend off thrown remarks, questions, and appeals for support in factional disputes. Each record, as selected from the shelves, represents an epiphoric instant of the shaping of interpretative knowledge. And each translates soon into a chain of enigmatic grunts and maybe conversations and purchases.

Generally ill-educated in the repertoire of classical music and the history of its interpretations and record-ings, for me this place and its procedures represent a climate of learning. In any case, my interest in vinyl is more for the singularity of each piece of tattered plastic than it is for what it represents more abstractly as the tokening of a history of performance or manufacture. With regard to this, however, the shop's commercial pol-icy also represents a preservationist strategy. Many similar commercial concerns may get hold of a collection for dispersal, fillet it for the valuable materials, and quite literally throw the remainder away. Much becomes lost to many. Here, at Gramex, records are sold at a more or less uniform and cheap price. For a pound, it is possible to pick up something that may be worth much more—financially. That is not what is said to be treasured, however. What is preserved is an archive of recordings. It is an archive that circulates in and out of the shop, much of it returning, accumulating price stickers. And perhaps what is more important is the fact that it maintains and preserves the possibilities for cultures and milieus and acquaintances and friendships to develop. This is an economic formation, with cultural implications. It is an interpellating fraction of capitalism, with its own rewards.

I spoke of a kind of commercial somatic clumsiness with reference to the Smithson's understandings of arcades. The clientele of Gramex have mythical status. Although as individuals they are not especially known for their professions or their domestic or sexual arrangements, they are certainly these too. In fact, as clients they are revered barely at all as individuals, but rather for their collections and the specificities of their expertise. In short, they are appreciated for their ability to forestall embarrassment.

Each has a kind of figural status, some not as heroic as one might hope. There is the case, apocryphal possibly but repeated nevertheless, of the learned though greatly disliked customer whose extensive collection was sold at the shop after his death. It was deliberately sold off very cheaply, for pence, utterly ignoring the commercially pragmatic probities of record dealing. It was an obituary humiliation. This anecdotal trace indicates the fact that, at all times in this place, evidence is sought as to the nature, conditions, development—the *order* of one's interests. Though no music plays, the space of this shop is one utterly structured by listeners and listening. Nothing is missed or passed over without observers' making some kind of spontaneously improvised sense of it. It takes time to appreciate that every remark, every flippancy is listened to, and lent weight to. Once, I made a few conversational gambits by asking innocently about pianist Isabelle Vengerova, about whom I knew nothing. I found myself a few days later being earnestly quizzed about other pianists, about whom I also knew nothing. A clumsy remark dropped, out of context and in company, precipitated a series of expectations, and all the embarrassments likely to ensue in front of a learned, listening audience. Here it starts to become evident that the relationships between masochism and sadism identified by Diller and Scofidio and Klein take on a powerful social form.

These figures, these walking, talking representatives of their collections, and their collections' representations of conventions and idiosyncrasies in the history of recorded musical culture, are not as stable as these anecdotes may make things seem. The delight is in watching and listening to the scale of the quarrels between them. Epiphora burst in front of you when wondering quite what it might be that someone makes of the significance of this or that record, which he has just slipped insouciantly back onto the shelf, miles from where it was found. Here the record, held only momentarily in pause, functions as an interpellative token. It is always potentially embarrassing when someone asks why you put this or that one back. This is another important feature of the shop. The records are not sorted in any way. It is as if they seek the state of the boy's bookshelves in Collete's story. There is no singular taxonomy imposed. The records go on the shelves, anywhere. The CDs are merely heaped. Archival stability is impossible there; it is impossible to find anything quickly, should you put it down. The place demands a physicality of constant leafing through the accumulation of piles of records as one browses, returning them to the nearest patch of empty shelving. What develop on the shelves are tiny orderings, little collections representative of something, certainly, but quite what is represented is exactly imperceptible, exactly epiphoric. What may be found, then, on the shelves are transient traces of acts of archival microresistance to HMV, Virgin, or Tower Records, their taxonomies and the pointlessness of staff returning stock to where it should be, and where the only way of finding a similar form of embarrassing association is to present oneself at the sales counter to play Twenty Questions: "It has a blue cover … they walked off Top of the Pops … it was on the radio last night."

———

The major point here, concerning the most cherishable aspect of the audition of Gramex is this: one hears the clientele move and shuffle and grunt and chatter and hold forth, but internally one may also hear the pleasantly cacophonous sound of each's collection, and the figural discordance and euphonies of them together. In this context, the tiny orderings on Gramex's shelves themselves then further suggest forms of microresistance. This resistance is not merely to the reified historical and generic categories that may be suggested by a commercial cataloging system. It is not even, at a remove, resistance to the ways that electronic databases may recognize a customer as one of particular tastes and tailor email advertising to them accordingly. Such archival groupings also register a social practice of resistance to the very notion of a well-behaved collection.

———

One afternoon, I had been in Gramex for some time, leafing through LPs, occasionally breaking off to chat, and from a rather large pile that I had accumulated, I selected only a few. I handed them over to David, who runs the till, for him to price up. I remember this too vividly. There was a recording of a Schubert sonata by Vladimir Sofronitsky, and one of the same piece played by Daniel Barenboim. There was a less-known Schubert opera, and another by Wagner, *Die Feen*. There were some things played by Clara Haskil and a recital by Dinu Lipatti—all in all, perhaps amounting to five or six pounds. David went through them, his face chewing in arch disdain. He has his own enthusiasms, for Bax and Vaughan Williams. At exactly that moment, one of the other characters in the shop, who was waiting and watching the transaction, said in a pronounced Austrian accent, "Such good taste. And so cheap. This is almost free." In that instant, possibly because three or four others had perked up at this point and were listening, he blushed and I blushed and David blushed. Exactly why, I am not certain, though we could speculate about uncertainties, and rankings and ranklings, personal inadequacies, or just a fear of saying the wrong thing. All of these flashed and squabbled in the redness, but to decide on one or a collaboration of a few causes would amount to something like successive speech, and the loss of the epiphora of embarrassment.

Melanie Klein's account of the sadistically possessive attack on an archive of symbolic (talkative) domestic objects was figured also as an attack on a maternal figure—which both represented and was represented by that archive. The circulating archive of Gramex is in a sense threatening to those who are interested in it, for the very reason that it is circulating. Things, objects that may be simultaneously of both general and idiosyncratic interest may be lost to others quicker off the mark, wittier in the recognition of the values their collections place on their missing parts, for instance. The archive articulates threat because it is also the medium of sustenance for a cultural micro-ecology. Possession may become unbearably important and profoundly disappointing. Klein's connection of this maternally figured archive with an essential disbelief (itself sadistic) in the authenticity of others' interests in the archive, and what it variously represents, could clarify something of what uncomfortably flashed forward at David's till. Certainly, it is somewhere here that the architecture of Gramex resides. So, the conversation turned quickly to a small sum of money, and I trotted off.

6.1　Wax portrait of conductor Herbert von Karajan. © Miracles Wax Museum, Salzburg.

6 Waxing on Walls

Yet, though I should have felt I had received less than my money's
worth had I travelled to Bavaria to hear this performance, it was
sufficiently good to be a valuable study in one's home with the text
open and one's imagination and memory to supply what was wanting.

—Dyneley Hussey[1]

Here architect Bernard Tschumi and conductor Herbert von Karajan meet to find similarities between the graphic line in recorded music and the graphic line in polemical architecture. We'll look at the line taken by different operatic coiffures. And, listening to Wagner, we'll see how the line drawn through Germany in 1952 sutures together national historical spaces.

1 Herbert's Haircut

"Where everything you want belongs to someone else":[2] lyrical, it is like a catchphrase. One of Bernard Tschumi's. Any rehearsal of the general familiarity with Tschumi's architectural procedures will highlight his habit of borrowing. This is especially so in his articulation of the ungraspable qualities of the *event*. That rehearsal would also admit that while helping to secure the idea of the event as an interruption in the authority of *function* for architectural thought in the last decades of the twentieth century, Tschumi also demonstrated a reserved and nostalgic interest in a well-established vein of dark urban humor. Evidenced by his teaching of Franz Kafka's short story *The Burrow* as a way of introducing the agonizingly forked situation of a *tinnitally* apprehended subject of architecture, Tschumi's sometimes mirthless humor is also revealed through his borrowing of visual gestures made by Orson Welles, Luis Buñuel and Alfred Hitchcock, among others, in his project *The Manhattan Transcripts*.[3] What should be said at the outset is that this humor in Tschumi's work itself approaches a perception of lived urban politics.

Tschumi's liking for modes of black civic comedy had a kind of local precedent. This appeared in the form of a popular commercial recording of Richard Wagner's music drama *Die Meistersinger von Nürnberg*. On its release by EMI in 1972, it was well received as something of a landmark in the history of recorded music. Greatly appreciated in some circles for a variety of reasons, it was greatly despised in others. Often this was for the very same reasons, principally to do with the ubiquity of the sonorities hawked by the work's conductor, Herbert von Karajan (see figure 6.1).

Though the recording circulated in its famous way right at the threshold of the formulation of some of Tschumi's most influential early projects, it didn't formally involve Tschumi in any conventional way. The texts themselves do, however, involve each other profitably in terms of a notion of linearity for architectural and musical perspective and the way this might form an understanding of some of the surprising ways that national cultural spaces were sutured together in the Cold War era.

———

It is probably best to start with a specific though partial image of Bernard Tschumi. For the depiction of the otherwise architecturally imperceptible, Tschumi's use of the term *notation* remains an entirely convincing contribution. In *The Manhattan Transcripts*, Tschumi employed this term for the first time to stage dismay at the imperious hypostases invited to architectural perception by functionalist ideologies: their desire to sort and to be certain. In *The Transcripts* the device of the figure of the acrobat, paused photographically in trajectory or in momentary balance, is presented as the antithesis of control. The acrobat is used to establish a tussle between that which is arrested by the conventions of architectural depiction and that which is endlessly transformed by the *inhabitation* of architectural space.

In the later passages of *The Transcripts*, the toppling acrobat serves as a contracted simile for the production of architecture exactly as an attention to the vernacular, adrenalized, irregular, self-licensing, and unpredictably changing usage of architectural space that is heralded by its inhabitation. Tschumi's use of such figures as a specialized notational form for architecture is now familiar. With them, he indicated in *The Transcripts* a set of urban routes of traverse, intersection, and escape, each of which was sympathetically pedestrian, and each of which was sponsored by a sordid and desperately trivial *crime divers*—whimsically here, a murder. In reprising the relationship of the frozen and determining instant to that of the durational fluidity of a variously observable and narratively diverging whole, Tschumi paralleled his notations allegorically with finely muted renderings of fireworks. Cooler in emotional temperature, he augmented these metonymic figurations of the festal city with an epithet on the ability of fireworks to reflect on a proposed ontological condition of architecture—that is to say, an architecture "burned in vain," illuminating, constellating, and, as a figure of capitalist spectacle, at once diachronous and synchronous. His remarks here have since accrued the kind of gravity hitherto reserved for Henry Wotton's line on *commoditie*, or, more significantly perhaps, Le Corbusier's remarks on the masterly play of architectural forms "brought together in light."

Light, musical light, architectural light in the sense of its cinematic manipulation—all pervade the work done by *The Transcripts*. Especially, it is the fragmentary light inferred by the instant caught by the cinematic still in its volatile relation to the duration of an unfolding cinematic diegesis that figures Tschumi's proposal of a novel interpretation of an architectural demeanor founded in the *event*. This incidental apprehension of architecture refuses to defer to monothematic, integrationist narratives proposed by the aesthetic and organizational principles of functionalism. At the time he made it, Tschumi's gesture of combing cinema for architecture recalibrated approaches to architectural narrative.

The section titled "The Block," the dénouement of *The Transcripts*, brings his different cinematic references together through the deployment of visual leitmotifs. One of these devices in particular improvises, through a cumulative visual sequence and spatial syntax, a cinematic meaning for the kind of chirpy neoclassical proscenium arch that came to be one of the most irritatingly regular features of civic and commercial architecture in the late 1970s and early 1980s. In this graphic passage, through a number of drawn transformations, the decorative details of this arch start to be indistinguishable from the sprocket holes of celluloid film, especially when it is seen to be perched upon by some gigantically mythic, Harryhausen-esque bird. More important in these drawings, however, is the development of a male romantic support that laces Tschumi's architectural view with a crucial problematization of the cinematic auteurship of urban architectural space. This is where we see Tschumi borrowing the habits and signatures of others' conceptual personae. While Tschumi informally invited several cineastes to animate his architectural mise en scène, it is the name of Alfred Hitchcock with which he is most associated in this.

The acceptance of the idea of a murder as the civic architectural act par excellence is not new as an understanding. In his essay on Joseph Mankiewicz's 1952 film *Julius Caesar*, Roland Barthes connected the classical thought of murder for the civic good with the speculative political epistemologies that are made manifest in raked and unkempt hair.[4] Through a dozen formal allusions, though, it is Hitchcock who appears at this comic point in Tschumi's imagery of urban violence. This choice of auterial reference has its own political implications. In adopting Hitchcockian images as the *graphemes* of an architectural manifesto, what Tschumi delivers here does not equal Hitchcock's bubbling pleasure in the distribution of guilt, sexual anxiety, and larceny, or his glee in the punishment of all forms of minor vanity, incompetence, and impropriety. Rather, it

6.2
6.3
6.4   Bernard Tschumi, from the section
      "The Block" in *The Manhattan
      Transcripts* (pub. 1979). Courtesy
      © Bernard Tschumi.

both celebrates and parodies Hitchcock's commercially appealing cruelty as a form of urban knowledge. The suggestion of Hitchcock that hangs about these drawings approves a view that, as architectural qualities, such cruel and entertaining humiliations may constitute fundamental conditions of metropolitan civility, as well of national esteem and self-reflection.

The male support figured in this passage of *The Transcripts* carries traces of Hitchcockian ambiguously. The thickly approximate manner employed in Tschumi's graphic line helps claim a degree of play in this. The liberty afforded by graphic *width* is a recognized polemical device in the repertoire of architectural rendering. It is vivid in the kinds of approximation that Le Corbusier set to work when making allusions between antique and modern proportion in the section on the regulating line in *Vers une architecture*, for instance.

6.5   Le Corbusier, original amended post
card of Norte Dame from which the more
familiar detail in *Vers Une Architecture*
is taken. FLC L5(6)113, © FLC/DACS.

6.6a Bernard Tschumi, detail from the
section "The Block" in *The Manhattan
Transcripts* (pub. 1979). Courtesy
© Bernard Tschumi.

6.6b Elvis Presley, ca. 1953. Used
by permission, Elvis Presley
Enterprises, Inc.

For Tschumi, the glossily flopping haircut sported by his urban character here becomes a broad-limned architectural detail. It makes a suggestive conflation of references to any of Hitchcock's severally compromised men. James Mason or Rod Taylor may be seen, or Herbert Marshall or Ivor Novello—even Peter Lorre. Caught romantically in what appears to be a death throe, this suited figure also brings an acoustic imagination to Tschumi's architectural argument. The sprocket holes announce the acoustic signature of the mechanism of the film projector. But there are points where those sprocket holes have become bullet wounds. The punctuating crack of a pistol-shot stands in, then, for the recalled affective tensions of Bernard Herrmann's scoring techniques. However they appear, special aural synchronies are lent to architectural space. At this point, the convulsion of this dubiously cartooned heroic figure is not entirely explicit or insistent in its connotations. That may well be a handgun seen trickling from his straitened mortal fingers. However, the deject glamour drawn erotically into the stricken and crumpling innocence of this criminal body could, at the same time, suggest the grammar of generic postures struck around such an iconic aural and visual image as the Shure 55c microphone: the "Big Elvis."

The Big Elvis microphone is an icon with a rather different set of cultural connotations to those other and similarly mythologized microphones of the period: the Neumann M-50 or U-47, which we'll return to.[5] For Tschumi, then, our figure may just be Elvis, or any of the number of similarly styled vocal tragedians that followed. Ian McCulloch, who sang with '80s psychedelics Echo and the Bunnymen, might be seen as such an icon, or Martin Fry, who was the front man for ABC. There may even be a hint of Joe Strummer's period quiff. In any case, that architectural hairline that Tschumi brings to bear on urbanity presents an image of collapse that connects Elvis's musical libidinal pensiveness with the impression of a just, blazing, and agonizing end.

As a means of figuring vernacular romanticism for architectural perception, Tschumi's graphic line appears complex and historically various in the architectural and pop-iconic referents it may suture. One thing is tolerably certain: this line, particularly this hairline, was probably not expected to have as a connotation the image of the Austrian conductor Herbert von Karajan. This is despite the fact that, as can be seen, Karajan's was a striking coiffure, appearing to inhabit a common and popular poetic fluency not just with Elvis but also others. Later we'll see that German conductor Hans Knappertsbusch could figure in this minor aspect of traditional musical iconography, or Hungarian Arthur Nikisch.

The introduction of Karajan to Tschumi at this point exploits a possibility. While Hitchcock may be present in Tschumi's early urbanism, as an epistemological footnote or temporarily vantage of address, it may be that he is *not* evoked to the extent of declaring a shared methodological structure; even if that identification was

6.7 Innovative Neumann microphones radically transformed the nature of acoustically apprehended space during the 1950s. The U-47 (left) and the M-50 (right). Courtesy Georg Neumann GmbH.

actively wished for by Tschumi. It is true that Tschumi's cinematic auteurs are not meant as some kind of theoretically applied decoration. Neither, though, does any one of them exemplify a single point of methodological gravity for the findings of *The Transcripts*. This is because in terms of a commentary on the means of architecture's achievement of vernacular spatial narrative, what *The Transcripts* offers is the tableau structure of the cinematic storyboard itself, and not any exemplary authorial usage of it.

Tschumi took care with this formal matter. In a pedagogic introduction to the conceptual conceits employed by *The Transcripts*, he provided a table of approved procedures for manufacturing appropriately characterized spatial propinquities. He recommended devices derived from cinematic editing techniques, such as repetitive, disjunctive, distorted, fade-in and insertive sequences, as they have been used by his cinematic heroes. In this *tabulatur* of cinematic possibilities for the editing of architectural space, however, Tschumi especially celebrated the ambiguities opened onto by the unexpectedness of the jump-cut, for the creative speculation, ambiguity, and surprise it provokeas. Through these different editorial techniques of dramatic spatial revelation, Tschumi proffers a hope that architecture will be realized as a carefully edited, narrative medium of surprise, one that announces new modes for the representation and effecting of both architectural and subjective transition.

6.8   Herbert von Karajan as photographed
       by Siegfried Lauterwasser for the liner
       notes of EMI's 1972 release of *Die Meis-*
       tersinger von Nürnberg. Photograph ©
       Siegfried Lauterwasser.

Tschumi's relationship to Hitchcock is essentially one of pragmatic and nostalgic disposal. If only in terms of shared compositional means, Tschumi has much more in common with Richard Wagner. Tschumi's often-made remarks on a desire for an architecture of "love and death" may already signal a Wagnerian senti-ment. The parallels between these two characters, each involved in their respective ways with a reenvisioning of architectural drama, are perhaps clarified most in Wagner's dark and cruel comedy *Die Meistersinger von Nürnberg*.[6]

A representative vestige of eighteenth- and nineteenth-century images of the festal city as it is sung, *Die Meistersinger* is a work of overt civic commentary. Moreover, in the 1970s diverse popular forms ranging from the cheery sunshine of Burt Bacharach to the darker lights of Joy Division reprised similar forms of musical commentary on urban sensibility. This latter temporal context is important because the similarities between *The Transcripts* and *Die Meistersinger* lay principally in the former's affinities toward the technical and poetic details in, specifically, the studio recording of *Die Meistersinger* that was conducted by Herbert von Karajan in the 1970s.

Made in a joint enterprise between EMI and the East German state-owned company VEB Deutsche Schallplatten, this particular rendering of Wagner's late work drew on the talents and reputations of singers Theo Adam, Geraint Evans, and Helen Donath, as well as the members of the Dresdner Staatskapelle orchestra.[7] Commencing in November 1970, this stereo recording of Wagner's most musically debt-laden work was made in a series of sessions in Georg Weidenbach's war-damaged and recently rebuilt Lukaskirche in Dresden. Among other promotional gestures, the release of the recording provided the occasion for Karajan to famously repeat his agent's observation that the playing of the Staatskapelle "shines like old gold."[8] It is in both the soundplan technically contrived by editing the fragmented performance of the work made in these studio sessions at the Lukaskirche and the cognitive expectations and changed aesthetic preferences of a commercially imagined, listening subject that was anticipated by the completed recording, where aspects of the character of *The Transcripts* appear to be predicted.

Wagner and Tschumi share certain devices. The finger-twisting humor, the exercise of the cumulating leitmotif, and the eruption into easy violence of an apparently genial urban sociability are common to both. But there are also connections between von Karajan and Tschumi. As a spatial technician whose instruments, as Glenn Gould remarked, included the sound engineering staff as much as the baton, choir, and orchestra, Herbert von Karajan contributed to a radical rethinking of what constituted a plausibility in the aural representation of space during his career. Much of this rethinking was articulated by this EMI recording.

This was a politically complex production, too. It involved producers and sound engineers from both EMI (Ronald Kinloch Anderson and Christopher Parker) and VEB Deutsche Schallplatten (Diether Gerhardt Worm and Klaus Strüben) who were unfamiliar with each other. None of them could be said to be regular studio colleagues of Karajan, either. This recording required state diplomacies as well, and required them at a moment of political possibility, not just between East and West Germany, but also

6.9 Georg Weidenbach, Lukaskirche, Dresden (1899–1903), photographed in 1970. Photograph © Siegfried Lauterwasser.

between the respective international powers with which they each established their often troubled alignments. The aural representation of what happened in the Lukaskirche then was made and informed as much by these other cadres and individuals as it was with the more usually celebrated musical participants—Adam, Evans, Donath, the Staatskapelle, and others.

Glenn Gould's appreciation of Karajan is important here for the suggestion of an apparent change in the representation of musical events that was fostered by shifts in the aesthetic habits of engineering teams in the period from the early 1950s to the early 1970s. By 1970, Gould had already argued that a movement was perceptible in Karajan's recordings. It was one that signaled a lessening determination to provide for the listener the spatially perspectival "evocation of a concert experience" in favor of a practice that "subscribes to that philosophy of recording which admits the futility of emulating concert hall sonorities by the deliberate limitation of studio techniques."[9]

Gould's distaste for the hobbling of studio techniques in the implausible pursuit of a single-point aural perspective, and his view of the deliberate reassessment of aural-spatial possibilities through the furthering of what may be possible in the studio, was nowhere near representative of all commercial recording practices in the twenty or so years after the end of the Second World War. The London-based imprimatur Westminster, for instance, continued to make "natural balance" monophonic recordings with a single, judiciously placed microphone well into the 1960s, with the aim of fostering an aesthetic pleasure by making a convincing representation of the space in which the performance occurred. Their recordings, some of which have been recently revived in vinyl formats, are still fondly recalled; some are celebrated as masterpieces of the art. Moreover, by the early 1970s even the technical possibilities of stereo had led to differences over the probities of the acoustic representation of space. These ranged between those spaces found by the ascetic "minimal-pair" procedures adopted by RCA and Mercury, for example, and the theatrical fancies invented by Decca, and especially in the recording of Georg Solti's remarkably popular recording of Wagner's Ring tetralogy, from 1958 onward.

It is also true that in that short period after the advent of the long-playing record (ca. 1952) and before 1958 (the moment when new disk-cutting technologies enabled the pervasiveness of stereo), a complex aural-spatial vocabulary was developed both by recordings themselves and by music critics remarking upon them. It was a vocabulary that was, it must be said, largely indifferent to the conceits of the single-point aural-architectural perspective. In such critical writing as that offered in the inventive language of critics Edward Sackville-West and Desmond Shawe-Taylor, for instance, we find a spatial poetics marked not by any single characteristic but by a variousness. Sackville-West and Shawe-Taylor were of a type of critic at the time who, rather than confining themselves to remarks on the history of a piece, or the interpretation of it documented by a particular release, felt licensed to speak sensibly and determiningly of the *qualia* of a recording.[10]

Their language is instructive. Music that is "blurred" by the studio acoustic, "like a watercolour left out in the rain," is an exemplary response to simple technical insufficiency. Sackville-West and Shawe-Taylor could worry about sections of an orchestra lost to a recording or complain about surface noise from a poor pressing like anyone else. More significantly, a general tone of playing captured in a recording might for them be "pleasantly forward, clean and agreeably sinewy." It may be tubby or wiry, dry or cheap. It might be intelligent. It may even be grateful. There is an ethics of acoustically apprehended space in the imagery of Sackville-West and Shawe-Taylor that, given their standing and reception as critics, seems to have been widely understood, if not always agreed with. What their subjective poetic grammar of aural space did not bother to fetishize was a conservatively perspectival spatial reproduction of a soundstage.[11] Given the technical circumstances of the monophony of the commercial recordings that they inhabited during the mid-1950s, this cannot be surprising. Though for them there may be forms of depth to a soundstage, these could not be of the specific kinds that stereo was soon to make available.

The comments of Shawe-Taylor and Sackville-West are cited here not so much to suggest the idea that the advent of stereo recording represents some kind of further, Adornoan, technically effected strain of regressive listening. This is despite the fact that the sensationalist bait offered by Decca's early *Adventures in Stereo* recordings went a great way toward sponsoring the view that what was primarily to be gleaned and aesthetically appreciated from stereo was a specifically calibrated, rhetorically mimetic, spatial perspective. Rather, the purpose is to indicate that, during the period from the end of the Second World War into the very early years of the Cold War, the perception of acoustically recorded space, as occasioned by musical performance, had not relied on a restrictive, and indeed fallible, generally comprehended and imaginable analogy to stereoscopic vision, but derived a spatial metaphorics from a complex and fluxing monophony—not so very different, in its allusive capacities, from the function of Tschumi's unshaded graphic line.

2 Occasional Events

Recorded opera, or music drama, produces a coincidence of different apprehensions of space. Opera has its own monumentally dedicated buildings, with their own special spatial-acoustic as well as decorative complexities, and their own local cultural traditions of preferred repertoire and performers. The temporal and geographic places indicated by set and costume design, the psycho-dramatic spaces inhabited by the musical agonists, the acoustic and other characteristics of the concert hall or the recording studio, even the places in which recordings of such music are later heard, on gramophones in parlors or bedsits, on personal stereos on buses and at bus stops—all of these contribute to an extended and complex spatial faceting for opera.

Ancillary spaces, such as the offices and studies of historians and critics, or those patrons who supply for opera "the protection of civil law and regular royalties," are also implied in this extended space, often to decisive extent.[12] In the case of Wagner, the entire apparatus of Bayreuth may be connoted. This includes the workshops and rehearsal rooms where the technical competences of idealized forms of Wagnerian staging and performance have been the object of public controversy since the founding of the summer festival in 1876. But Bayreuth isn't the only theater; there is Munich or Covent Garden, for instance, and other recording studios, too. The types of microphone deployed and directed in each case, their proximities and directional sensitivities, how their signals are technically processed, the ways in which what they apprehend is blended, added to, or filtered at later stages: these things also clearly suggest the ways in which recordings of opera are caught in a web of surprising aesthetic, pedagogical, technical, historical, geographic, and spatial complexity.

The appreciation of *Die Meistersinger* is also susceptible to occasion. There is something festively inaugural to the history of the work. For instance, at the same time Marcel Journet was making an "acoustic" recording of Hans Sachs's *Wahn* monologue in Italian for the Gramophone Company in 1924, and Albert Coates was leading a long, English-language recording of excerpts of the piece, Fritz Busch conducted the performance of *Die Meistersinger* that opened the first of the postwar reinaugurations of the festival. This was the moment when the Bayreuth festival was rehabilitated after the First World War. It was also the moment when the foundling National Socialist German Workers Party adopted Bayreuth as its spiritual home.

The festival underwent a postwar reinauguration in 1952. Like others, in reviewing the *Die Meistersinger* as the opening work that year, critic Adolf Aber expressed his reservations about new production values in the festival overall. Deploring what he saw as the inadequate, rather than preferably absent, lighting of Wieland Wagner's spare settings of the Ring cycle, he applauded the quiet disinclination of the audience to burst into the spontaneous choruses of "Deutschland, Deutschland über alles," with an emphasis on the imperial rather than unificatory sentiment, that had marked the end of the performance of *Die Meistersinger* conducted by Busch in 1924.[13]

It is yet possible that a recording of Busch's 1924 Bayreuth performance may surface. Such things do keep appearing. Portable acoustic recording techniques, though far from refined, had already been used outside the studio to capture crowd noises and other fugitive yet marketable features of modern urbanity.[14] A possibility: it hasn't been found to date. So, there is an acoustic void represented in this kind of aural history of Bayreuth; monumental, and approachable only infinitesimally, by anecdotes.

No darling of the Nazis, Fritz Busch in time discovered the fleeting value of his Aryan credentials, no matter how impeccable. To preempt an SA-staged riot, he left the stage just moments before starting a performance of *Rigoletto* in Dresden. For his vocal antagonism toward Nazism, and perhaps FOR his family ties with the Jewish pianist Rudolf Serkin, he was forced to stand down from his position as director of the opera house at Dresden, and eventually to leave Germany altogether.[15] In 1933 at Bayreuth, a further historiographic highlight emerged in the form of an astonishing performance of *Die Meistersinger* directed and designed by Emil Preetorius and Heinz Tietjen, with a chorus numbering eight hundred thought necessary to produce an appropriate, communally martial flavor for the merrily ritualized humiliation of the character of Sixtus Beckmesser that concludes the work.[16] *Die Meistersinger* wasn't to be presented again at Bayreuth until 1943, this time under the baton of Wilhelm Furtwängler. By this point Berlin and Munich had become the state-preferred venues for the performance of the work in the intervening period. Even this 1943 performance was a staging directed not for a German bourgeois audience but rather for a group of munitions workers.

In 1944 another archival hole appears. Helmut Krüger and Eva Derenburg, sound engineers with the Reichs-Rundfunk-Gesellschaft, produced a stereo recording of a live performance of *Die Meistersinger* at Bayreuth. As with the other many stereo recordings of classical repertoire made in Nazi Germany before the availability of domestic stereo record players, the tapes of these performances are now lost. Even in absentia, then, it seems that occasional recordings of *Die Meistersinger* are capable of reminding us of an awkward past.

And so it goes. From 1924, to date and retrospectively, at all points, and often with good enough reason, *Die Meistersinger von Nürnberg* has been broadly judged in the critical literature in term of the burden of both the fractured political narratives of Nazism in German culture and Richard Wagner's apparently prescient sympathies for them. Either *Die Meistersinger* is thought to exemplify Nazi ideological traits, or in each production it is caught in more or less successful attempts at the recuperation of itself and its author from them. These themes form the fabric of a refrain for much critical commentary that is regularly and interestingly improvised upon.[17] Against this critical ground, however, the general tenor of the *political* performances around the Bayreuth festivals from the early 1950s to the early 1970s has also been caught up in a struggle to radically renew the relevance of Wagner's music, and to make aesthetically convincing arguments for the continued place for his works in the operatic repertoire. This continuing stain on *Die Meistersinger* insists on a symptomatic rather than what we'll come to see as a *sinthomatic* character for this most populist of Wagner's mature compositions.[18]

———

The earliest full recording of *Die Meistersinger* at Bayreuth was coincidentally also conducted by Herbert von Karajan for EMI/Columbia. Made available early in 1952, it came as a documentary epiphenomenon of another post–World War II rehabilitation of the work, the 1951 festival. Edited together from several recorded concert performances, it was released initially as a bundle of thirty-six, 78 rpm shellac disks.

This represented quite a commitment of space to the pursuit of musical culture for any home. The act of subjective endurance countenanced by listening to a four-hour opera in four-minute sections has its own ramifications for a history of aural attention. Released again that same year in the more compact form of five 12-inch LPs, which were pressed to allow for use with new automatic disk-changing facilities on gramophone reproducers, this first complete recording of *Die Meistersinger* with Karajan is one that marks a point of considerable transition in the architectural and temporal impact of recorded sound on domestic space. With this change, an easy repeatability of long pieces meant that such music started to become repeatable, familiar, and domestically unobtrusive—an option for the background to domestic chores.[19]

Only two years after this release, in 1954, the preliminary commercial experiments with stereo were to start, and things began to change. The conventions of stereo spatialization became standardized over time, enabled

6.10 Hans Reissinger, stage set of act 2 of
Die Meistersinger at Bayreuth in 1933.
Photograph: A. Pieperhof. Courtesy
© Bayreuther Festspiele GmbH.

not just by advances in cutting-edge technology, as mentioned already, but by new microphone setups, such as those eventually agreed and adopted by Decca, French Radio (ORTF), RCA, Mercury, and others, each improvising on techniques devised and heralded by engineers like Alan Blumlein, Arthur Haddy, and later Jürg Jecklin.[20] To the context of an emergent history of stereo-spatial standardization, it is worth adding that the short period of the production of monoaural LPs in the 1950s is interesting precisely because of the lack of microphonic standardization. The spatial effects produced by this lack of standardization lent veracity to the poetic investigations of critics like Sackville-West and Shawe-Taylor. The spatial burden to which stereo recordings may be made critically accountable can be sternly uninventive, and in its most authoritarian forms may demand an oversimplified understanding of the model of Renaissance visual perspective, one lacking a knowledge of the mystical narratives frequently articulated by its achievement of space. This is, again as mentioned, despite the kinds of invention brought to the popular impression of Wagnerian space by Decca's Wagner recordings with Georg Solti after 1958. Solti's recordings were both loved and vilified for their aural dramaturgical effects, such as the way Mime is given spatial omnipresence as, invisible, he moves about scolding and poking his brother Alberich in *Das Rheingold*.

In its recorded forms, *Die Meistersinger* sits habitually at several thresholds in the rapidly changing historical landscape of aural-architectural poetics. These thresholds concern the domestic practicalities of modes of attention to recorded sound. As period bookends, the two Karajan recordings (from 1952 and 1972) are perhaps too easily marked in terms of their relevance to European political history. The unseemly enthusiasm of Karajan in joining the Nazi party twice during 1933, for instance, has been a repeating and embarrassing trope for his biographers. In 1952, this knowledge came to figure politely in the critical reviews of the release of EMI/Columbia's *Die Meistersinger*. These two recordings also carry a Cold War burden that seems just as difficult to shake. While signaling polar conceptions of the articulation of the space produced and poetically exercised by a performance (one in 1951, monophonic and live, and the other in 1972, stereo and studio bound), Karajan's two *Die Meistersingers* parallel other historical events in the conception of the Germany.

In March 1952, two months before the border between East and West Germany was closed, Joseph Stalin arranged for a note to be sent to the negotiators for the Allied Forces in Berlin. He suggested that since

Germany was so important to the stability of postwar Europe, all interested parties should settle their differences and collude to maintain a thriving and unified Germany. Stalin may have benefited greatly from this. Suspicions over Stalin's intentions here has produced one of the great conundrums for Cold War historiographers, and the question of what was meant by this gesture is unlikely to be settled.[21] In the period from 1969 to 1972, Egon Bahr was engaged in the conversations that led to the forms of rapprochement that were to enable Willi Brandt's *Ostpolitik* initiatives to come to something. And, it is difficult to imagine Karajan's journey across the Berlin Wall to Dresden, to engage in an artistic collaboration over one of the key texts of the meaning of Germany, without conjuring as a pretext Egon Bahr's surreptitious deliberations. Both of these are just two further historical circumstances that have shaped understandings of some of the very many commercially available recordings of *Die Meistersinger*.[22]

3 The Raked and the Unkempt

The hairline has a distinguished place in the history of critical remarks relating to the emotional tenor of Wagnerian performance. Regarding a diminution of Wagner's music to a comforting and unremarkable aural backdrop to a nobiliary and exclusive mode of cultured association, Claude Debussy noted of pianist-conductor Alfred Cortot in 1903 that he was one who has "most profited from the pantomime customary amongst German conductors. … He has Nikisch's lock of hair." Debussy described its fascination:

See how it falls, sad and weary, at moments of tenderness in such a way as to interrupt all communication between Monsieur Cortot and the orchestra; then see how it stands proudly on end at warlike moments.[23]

Concerning the focus on emotional atmosphere that has characterized many other passing remarks made about the kinds of experience occasioned by the performance of Wagner's music, it is worth isolating two specific strands of thought that lend something further to the subjectively reticulated possibilities of carefully raked hair. The first of these concerns the *illumination* of the audience. In a short essay, Romain Rolland described his introduction to Wagner, and the performances conducted by Jules Pasdeloup at the Circe d'Hiver in Paris:

I was taken there one dull and foggy Sunday afternoon; and as we left the yellow fog outside and entered the hall we were met by an overpowering warmth, a dazzling blaze of light, and the murmuring of the voice of a crowd. My eyes were blinded, I breathed with difficulty and my limbs soon became cramped; for we sat on wooden benches, crushed in a narrow space between solid walls of human beings. But with the first note of the music all was forgotten, and one fell into a state of painful yet delicious torpor.[24]

Toward the end of his essay, Rolland recounted a later conversation with his friend Malwida von Meysenbug. In 1876, she said, while staring through the Bayreuth darkness at the stage, two hands reached from behind her to cover her eyes. And there was Wagner's impatient voice: "Do not look so much at what is going on. Listen!" Sage advice, Rolland recommended, to listen to Wagner with eyes closed, but completely at odds with what seemed to preoccupy him in his essay. This is because, here, Rolland's most telling remarks have to do with the visual convening of community. Thrilled with pleasure and pain, invigorated, strengthened, gladdened by the music and its novelty, he wrote, it seemed as if "my child's heart were tore from me and the heart of a hero put in its place." Rolland spoke of the irrelevance to him of the poor acoustics of the hall, and of the incompetence of players that ruined musical design. Nevertheless, in an audience of "poor and commonplace people" with their "faces lined with the wear and tear of a life without interest or ideals," he nevertheless *saw* his own emotions reflected. This near-Baudelairean shifting of interest to ancillary social details, those of the audience, characteristic of much of Rolland's writing, is recognizable as a predominating strain of criticism. George Bernard Shaw is another who enjoyed lending critical decisiveness to the shoulder-rubbing foibles of an audience, as was Frederick Nixon. Writing, in 1952, of the same period that Rolland remembered, though at Covent Garden, Nixon said that "no person of taste would speak of seeing an opera, one saw a play, but one heard an opera." The appropriateness of the epithet was, he said, due to a lack of production values in any "modern," perhaps Wagnerian, sense.[25] No distractingly indelicate onstage physicality was allowed, and the music, as opposed to the "beautiful singing," he thought unremarkable. Vitally, for Nixon as for Rolland, the spectacle of opera was the audience itself.

Wagner's own aversion to the idea of opera as the occasion for a bourgeois audience to narcissistically admire its brilliantly lit self is well documented. He liked to reflect upon, police, and stage such intersubjectively unruly aspects of association within the constraints of a text like *Die Meistersinger*. This is a text that barely conceals itself as an allegory of bourgeois cultural habits. His disdain for the ordinary incontinences of an audience was one of his stated reasons for insisting on a murky darkness for the theater at Bayreuth. In some senses, the collapse of class distinctions, such as those noted by Rolland and Nixon, through a principled and conservative renewal of music, is the core proposal of *Die Meistersinger*. Ability shows where it will, whether it is in the guildsman brilliance of Hans Sachs or the vital noble novelty of Walther von Stolzing. Only a scheming or critical naivety could find in Wagner's critique of bourgeois mores a parable for renewed democracy, however.

In giving meaning to the convened irregularity of a visible and illuminated audience, Rolland and Nixon undercut some of the assumptions of vernacularity hedged by Wagner. In promoting the image of a collection of differently attentive listeners, and in resisting the image of an audience's communing self-sacrifice in terms of their personal comfort, Rolland and Nixon both align neatly with the vernacular imperatives of Tschumi's upsetting of the ordering requirements of architectural functionalism. Moreover, the physical substantiality of this illuminated audience has a particular architectural impact on the tuning of atmosphere for recorded performances.

4 Line and Light

The figural illumination of Wagner was something taken up elsewhere by Ernst Bloch. Bloch focused more on the philosophical work of *Die Meistersinger* than on the determining features of the rude politics caught up in the image of its audiences that had Rolland's attention. In a series of commentaries, scarcely ever departing from the poetic terrain of light as a means of figuring musical meaning, Bloch turned to the thwarted, prohibited, and unrequited loves that act as the vehicles for Wagner's other meditations on relationships between art and civility. Especially, Bloch attended to the way that the libidinously self-sacrificing cobbler, Sachs, characterized the dynamic of these relationships in his *Wahn* monologue. Bloch paralleled one distinction between line and light with another distinction: that between the indexical presence of a raw, irrepressible, eruptive, unsynthesized, gloriously vulgar and unconventional nature (*natura naturans*) and the depictions of it found in the figuration of a reflective, gestural, elegant, muscular, or dainty *poetics* of nature (*natura naturata*).

Resting on Schelling, Bloch suggested that whereas a composer, Bach for instance, may have been concerned with depicting what was geometrically fixed or frozen in a mathematical cosmogony, the "Romantics" fixed the desirable vulgarity of "*natura naturans* not as a diagram but as phosphorus."[26] This romantic light appears vividly in the *Wahn* monologue, and it does so as the form for the libidinal grounding of carelessly pretergressing manners of civic association that runs through *Die Meistersinger*. When Sachs performs the

Wahn monologue, this morbidly pessimistic consideration of the fevered state of social relationships of all kinds, Sach's mood often lightens fractionally. This is at the moment when he alludes to the relationship between Eva, the daughter of local, proto-bourgeois pillar, Veit Pogner, and the itinerant knight Walther von Stolzing, the character who represents a locus of Wagner's repining at the hobbling effects of a conservatively entrenched attachment to cultural tradition. At this moment, Sach apprehends these hero-lovers in the image of the flickering light of a pair of glowworms. They are irresistibly drawn to each other, and respond not to social trammeling but to other, more natural, less regularizable, more precarious, less civically affordable desires.

It is an image of love and death not unlike Tschumi's. It might push an analogy too far to say that Sachs's flickering glowworms, in the way they figure that snatch of timeless dark light Wagner developed in the Ring cycle, match the flickering fragments of cinematic light represented in Bernard Tschumi's endgame for *The Transcripts*. Nevertheless, this does draw out some important allusions for the architectural meaning of the thickness of Tschumi's graphic line in Elvis quiff. Bloch went on to make an overbroad historical observation. Music, he argued, entered the quadrivium of suitable humanist studies in the sixteenth and seventeenth centuries at some expense. This cost, he said, was a requirement to consider music spatially, as a mathematical function.[27] This is to say, as Bloch argued, the price of music's philosophical acceptance was that it should be frozen, mathematically, as a species of *natura naturata*: essentially modular, proportional and diagrammatic, and borne more by the perspectival techniques seen in Piero de la Francesca or Filippo Brunelleschi than that seen in, say, the flower paintings of a notable Wagnerite like Odilon Redon.[28] The lesson of Bloch, one seemingly understood by Tschumi, lies in Bloch's worries over the pretense of romantic art forms in the "copying of nature as the unearthing of nature." It is certain that it is only the phenomenal, cartoon superficies of Tschumi's drawings that exercise a Renaissance-derived, perspectival diagramatics. Otherwise, more like Redon, Tschumi's lines, in their critically allusive width, approach fluxes of phosphorescence.

———

A second strand of thought regarding the line concerns more abstract conceptions of the architectural structure of music. Of the most fluently put arguments about the formal architectonics of Wagner's works, those of Alfred Lorenz have held a pivotal position. It should be said that their importance derives greatly from the way they have been derogated by other critics. His analyses of the formal emotional unfolding of each of

Wagner's musical dramas return eventually to the powerfully abbreviated argument that he made about *Die Meistersinger*, for the occasion of the Bayreuth Festival of 1925. Arguing that, unlike any predecessor, Wagner was able to surpass the traditional operatic habit of linking a series of numbers (arias, ensembles, finales, etc.) with passages of recitative and dialogue, Lorenz proposed that Wagner had evidently been able to hypermetrically resolve the emotional narrative of each of his works in a single appropriate form, a form that "arises out of the innermost core of the work at the very moment of its artistic conception."[29] In the case of *Die Meistersinger* Lorenz suggested the architectural image of the arch for this form. The precedent for such a form may perhaps have been sponsored for Lorenz, in part, by the setting of the opening scenes of *Die Meistersinger* in St. Catherine's church in Nuremberg, incidentally indicating a Catholic cultural location for its narrative. Gravity-bound and geometrically simple, Lorenz's architectural view was perhaps forgivably constrained by loadbearing architectural forms, such as the Gothic, Romanesque, or other kinds of arch.

Such an image has worked well in providing a spatial analogue to the possible, architecturally descriptive aspects of Bruckner's music, but it is perhaps less apt for Wagner. Theodor Adorno disagreed with Lorenz. At least, in drawing attention to Lorenz, he articulated his own reservations regarding the notion of the simple predictive architecture for Wagner's music that is suggested in the apparent fidelity of the arch. Adorno did this by proposing a radical complexification of the emotional line taken by the work. Having already introduced the formal picture of the "wave" in an argument regarding the threshold of *limits* in Wagner's attempt to "invalidate anything with a definite shape, to make everything flow and obliterate every clear frontier,"[30] Adorno later said of *Die Meistersinger* that its "continuity is created, over long stretches, by an unconstrained redrawing of the dramatic curve, from moment to moment."[31]

With this figuration of a surprising, vertiginously emotional calculus for the appeal of Wagnerian linear structures, Adorno followed the specific mathematical disdain articulated by Bloch. At the same time, Adorno refused the proposed relation of the Wagnerian emotional *instant* to any atemporal predictability architecturally insisted upon by Lorenz. In this, Adorno handed to the listening habits of an audience a view of such features as the leitmotif, not as burden-bearing elements, predictable in their affects, but as infinitesimals, anecdotes. This view comes with a proviso that, insofar as the proximal character of such features approach, but never partake in a line, anecdotal experiences of Wagner's music also bring to the interpretation of the endlessly redirecting line of the music something of the diverting *remous* of civic association.

This cluster of partial theoretical observations on Wagner is arranged pragmatically. Here, the intersubjective variety of Rolland and Nixon's audiences, as they are bought together as a felicitous emotional plebiscite in a flux of light, is raked together with a series of reflections on the conventionally understood imprecations to the limiting and descriptive function of the musical *and* the architectural line. For Adorno's conception, it seems that the hairline, as variously deployed by both Tschumi and, as we'll see, Karajan, may provide the perfect analogue from which it is now possible to depart. More importantly, Adorno musically shapes a means via which it is possible to see the socio-architectural relationships between line and light in Tschumi's notation of the adrenalized content of unexpectedness for architectural narrative. In his very handling of the graphic line and in his decorative mobilization of motivic nostalgias gleaned from fractions of cinematic light, it can now be seen that Tschumi warrants an effective dislimning of one of architecture's representational means.[32]

5 Engineer's Aesthetic (Wo bist du?)

It may come about only through the exercise of familiarity but, nevertheless, it is possible to acquire a marked proficiency in identifying the microphonically construed spaces in which music is performed. On occasion, the acoustic signatures supplied by BBC Radio of the Wigmore Hall in London, for example, or the Ulster Hall in Belfast, St. David's in Cardiff, or the Saturday matinée broadcasts from the New York Metropolitan Opera House, are frequently much more recognizable than the music being performed. The apprehension, or rather the articulation of an architecturally characteristic ambience is, as we've seen, vital both to the aesthetic experience of recorded music and to its criticism. This is in terms of the pragmatics of containing unruly eigentonic effects of a venue, through the judicious placing or balancing of microphones, or in vitiating the deleterious effects of overlong resonances on the cognitive abilities of musicians' attentions to volume. Regarding this, in his influential technical manual on sound recording, Alec Nisbett described a procedure of ambiophony. Ambiophony is a method that uses a number of microphones and loudspeakers to produce for the performers the aural impression that they are playing in a smaller hall than the one they are actually in. This helps prevent a tendency to overdrive their playing to an imagined audience at the back of the hall, when theirs may in fact be a much more intimately imaginable audience of phonographic listeners.

For critics, a shared grammar of motifs to describe the sound of a space is regularly found in references to light. A luminousness or a radiance is one of the stakes in a positive review of a recording. The significance, here, is that as much as sound engineers and producers need to attend to the physical and intellectual conditions of a good performance, so too they must attend to the representation of the place of the performance for the recording. Acoustic dramaturge John Culshaw, the producer of Decca's early stereo cycle of Wagner's works with Georg Solti, recalled an incident in his autobiography involving Wilhelm Furtwängler. During a recording session at Kingsway Hall, Furtwängler insisted that just a single microphone be placed in the hall, emulating what he thought to be the techniques of the German engineers with whom he was more familiar. With no time to experiment with the placement of the microphone, the result of the session was an unpublishable acoustic disaster.

The poetics of architectural representation for sound recording is an enormously nuanced art, and despite various audiophiliac insistences on the superiority of one or another commercially adopted method, there is no single ideal way of recording a room, monophonically or stereophonically. This plurally creative dimension remains vital to the fabric of the meaning of any piece of recorded music. The narrative necessity of an acoustically convincing fiction of architectural place is dramatically revealed, for example, when it comes to digitally mixing the apprehended source materials in more recent multimicrophone recording techniques. These approaches, essentially monophonic, involve the placement of microphones close to each performer. The aim is to get, as far as is possible, a recording of the individual without any overspill of sound from neighboring players. Their sound may be panned into spatial position later, recreating, perhaps, the effect of a more conventionally gained stereo image. The costliness of the time of studios and performers makes this practice desirable, as it saves the effort of balancing the microphones to the resonances of the performers and venue. To overcome the lack of any larger spatial dynamics to the aggregated recording, it is then possible to wet this dry preliminary recording by adding the sound signatures of specific venues through convolution software. These are often provided as industry standards.

This technical development, financially driven, shows the degree to which commercial recording can be said both to value and to have lost architecture and, in the process, to have become profoundly nostalgic about it. As a fetish here, and as an instance, this elegiac possession of architecture reveals a further portal onto the political character of the aurally recorded space occasioned by *Die Meistersinger*.

6 Twisted Pairs

In 1951, the year that Karajan made his first full recording of *Die Meistersinger* with EMI, the duties for conducting it at Bayreuth fell equally to him and to Hans Knappertsbusch. Within the broader context of Wieland Wagner's postwar renovation of Bayreuth productions, the first stagings of *Die Meistersinger* sit awkwardly, like revenants. The productions of *Parsifal* and the *The Ring* tetralogy were, in 1951, subject to complete reassessment and radically abstracted from the traditions of Wagnerian dramaturgy. But, for the "New Bayreuth," the first *Die Meistersinger*—that work most overtly figured as an ideal of German, urban, vernacular community, the work with the most compelling inaugurative manner, and the one most tainted by its recent political associations—it was old hands, those of Rudolf Hartmann and Hans Reissinger, who controlled the production and staging. Like Hartmann and Reissinger, many of the figures involved in the second reinauguration of Bayreuth had compromised political histories. It is known that, like Karajan, conductors Clemens Krauss and Karl Böhm had done well while staying on under the Reich. Even the sublimely avuncular Hans Knappertsbusch, remembered fondly for, among other things, the way he eventually fell out with the regime, was, it has been argued, partly responsible for the forced exile of Thomas Mann.[33] And, in his discussions with the Allied administration in Germany, after the war, it was made clear to Wieland Wagner that the festival would not go ahead at all without the certain and explicit distancing of his mother from the project; the far-from-nonpolitical Winifred. It is worth noting that here we see something of the hyperbolic expressionism of the Great German Art exhibitions in the details of Hans Reissinger's set designs both for his 1933 staging and this first postwar Bayreuth production.

The Reissinger–Hartmann production was put on again in 1952, and it was to be the last *Die Meistersinger* to be seen at Bayreuth in its architecturally naturalistic manner.[34] Its next incarnation was in 1956, the *Meistersinger ohne Nürnberg*. Produced by Wieland Wagner, this version had marked visual differences. The recognizable, stylized architectural depiction of Nuremberg, familiar until the time of Reissinger, was reduced to a set of ahistorical natural ciphers—and this may have been a conscious reflection of the devastated condition of postwar Nuremberg. In the second act, for example, the set consisted only of symbolically blooming, suitably lit trees—elders for wisdom, lindens for ardency—which stood outside the houses of Hans Sachs and Veit Pogner, respectively, in keeping with Richard Wagner's instructions.

6.11   Hans Reissinger, stage set of act 2
       of *Die Meistersinger* at Bayreuth
       in 1951–52. Courtesy © Bayreuther
       Festspiele GmbH.

In 1952, Hans Knappertsbusch alone conducted *Die Meistersinger* at the festival. Apart from some swapping of roles and the notable absence of Erich Kunz and Elizabeth Schwarzkopf (who was replaced by Swiss soprano, Lisa Della Casa), the production was much the same as that of the previous year. As much as Knappertsbusch's performances from 1952 may seem to wear the accoutrements of the end of one aspect of the history of the performance of *Die Meistersinger*, especially in its de-Nazification, they also represent a form of beginning. So, before returning to Karajan's later EMI recording and its connections to Tschumi, we should pay some

6.12 Wieland Wagner, compellingly reduced
stage design for Act 2 of *Die Meisters-*
inger (*Die Meistersinger ohne Nürn-*
berg), Bayreuth, 1956. Courtesy
© Bayreuther Festspiele GmbH.

attention to the nature of this beginning, in the form of a lately released recording of Knappertsbusch's reading of *Die Meistersinger*, made and broadcast by Bavarian Radio in 1952.[35]

In some regards, the large part of Wagner's oeuvre is taken to represent an upsetting of at least one dimension of perspective: musically, in the deployment of chromatic compositional means to produce allusive evasions of a fixed tonal perspective. Historically, critics have taken this, along with Wagner's facility with more conventional devices, such as the layering of differently grouped voices as heard in the blithely spontaneous urban riot at the end of act 2 of *Die Meistersinger*, to represent either acute and neurotic vice (Cui, Berlioz,

Ruskin), or the means through which Wagner redeemed music from the perceived calamities of a steady degeneration of its forms (Liszt, Newman, and in his way, Adorno).

Other forms of perspective than tonal ones are at stake here. The careful tuning of its lightly plastered, wood-paneled walls and ceiling contribute to the celebrated acoustic familiarity of the theater at Bayreuth, when that architecture is brought to light and aurally detailed by music. That peculiar resonance is augmented by the role played by the cover over the orchestra pit, and by the further facts that the orchestra pit is mainly underneath the stage, that the stage house is peculiarly large, and that there is only one story to the auditorium. For a live recording like Knappertsbusch's, there are further aural-architectural facets. The coughing, the conversation and applause, the way the wicker seating seems to amplify the audience's ordinary physical adjustments to suggest a restlessness, and the acoustic absorptiveness of the very bodies of that audience all manifest themselves. For one preferred mode of fantastic attention to recorded opera, the delinquencies of these and other apparently extradiegetic, Rollandian motifs, such as the clattering of instruments, the rustle of shirtsleeves and costume, footfall and the crinkling of scores, should be intellectually deleted by the listener,

6.13 Hans Kanppertsbusch's lock, ca. 1952. Photograph: G. Jaeckl. Courtesy Deutsches Theatermuseum München, Archiv Rudolf Betz.

as an act of aesthetic felicity.[36] These aggregated indices of presence, which figure pronouncedly in Knappertsbusch's recording, partly constitute the kinds of *qualia* that the recording may occasion for its listeners—that is to say its *atmosphere*.

What all these little, acoustic *effets réels* also signify is an uncertainty as to the location of microphones, and what microphones themselves interrupt as well as sustain, in the apprehension of the performance. In the 1952 Bavarian Radio recording of Knappertsbusch's *Die Meistersinger*, for instance, it isn't possible to tell if the sound of turning pages arises from a member of the audience, from a player or a prompt, or from Knappertsbusch himself. The placing of microphones at Bayreuth was a commercially contested topic at this time, and one subject to a deal of surreptitiousness. Mythology suggests that it was permitted at Bayreuth in the early 1950s to array "opera-mice" at the edge of the stage and to place microphones under the lip of the cowl covering only the orchestra.

The cowl is important to the sound at Bayreuth. In keeping with Wagner's aural architectural aesthetics, it serves both to hide the musicians, to blend their sound with that of the singers, and to direct this sound upward to reflect off the ceiling.[37] In their recordings, as John Culshaw and others have related, Decca's engineers were able to secure markedly different spatial images of the Bayreuth auditorium, from those of EMI and Bavarian Radio, during these first two years of the Festival.[38] Arthur Haddy managed to sneak a favored microphone into the roof space. For his performance in 1951, Karajan wished to cut a hole in the cowl covering the orchestra pit, perhaps so that he could be seen. If he had had his way, this may have disturbed the spatializing function of this acoustic-architectural fitting. Karajan did certainly rearrange the seating of the musicians in 1951, and this may in part account for the spatial differences between the Knappertsbusch and Karajan recordings. It is also evident from the number of recordings made at Bayreuth in the early 1950s that the singers' voices were foregrounded over the music, and that this changed after 1955. This somewhat giganticist vocal foregrounding exists in the Knappertsbusch recording, too.

———

This issue of the aesthetically appropriate blending of orchestral sound at Bayreuth stems directly from the influence of Gottfried Semper in the design of the theater. The agonizing tale of theft, personal betrayal, and official subterfuge that describes the reasons for Semper not overseeing the building of the theater himself has

6.14 The orchestra pit at Bayreuth. Note the steep downward rake of the seating away from the conductor's desk; note also the cowl that hides the orchestra from the audience, while helping to blend the sound made by the orchestra. Note too the columnation of the auditorium. Photograph: Jörg Schulze. © Bayreuther Festspiele GmbH.

been recounted elsewhere.[39] The key idea of locating the orchestra below the stage is fully attributable to him, in his discussions of the renovation of orchestral sound with Wagner. As figures in a historiography, Semper and Wagner were close for long periods, especially during their insurrectionary activities as progressive nationalists in Dresden in 1848. They were both present at the barricade built outside Semper's house. Semper, in fact, demanded that this makeshift wall, swiftly assembled from carts and domestic furnishings, be dismantled and reconstructed to his exacting architectural specification. Their friendship, their mutually acrimonious parting, and the nominal nature of their eventual, resigned, and distant reconciliation are all parts of the mythology of Wagner's work. Keen to borrow the agencies of others where it suited him, Wagner, it seems from the historical picture of him, scarcely squandered aesthetic opportunities to capitalize on his poisoned relationships with former friends and allies. Yet, unlike in other cases, that of Eduard Hanslick for example, it is difficult to see where Semper figures in Wagner's oeuvre as an exploitable caricature—except architecturally.

Fully thought through as architecture, Semper's insurrectionist's wall, contingent both in geopolitical terms and in the sense of scavenged and donated domestic materials used to construct it, is emblematic of the comradeship, rancor, and diplomacy that passed between him and Wagner, as the social currency of Dresden's vigorously critical artistic milieux. It also reflects on the eventually makeshift construction of the theater at Bayreuth. In the end, this was built hurriedly, intended as a temporary structure, by Otto Bruckwald. Bruckwald derived his building from the plans that Wagner had purloined from Semper, in the full and happy knowledge that the assurances of state remuneration Wagner had passed to Semper would come to nothing. Though lightly built of resonant timber rather than the more substantial materials Semper had in mind, Bayreuth's happy and singular acoustic still derives directly from an act of technical cultural theft.

The theoretical understanding that Semper ascribed to the wall for architectural vocabularies, especially its symbolic impermanence, is pertinent here. His own idiosyncratic views of antique culture placed the temple at the center of ritual social activity, the convening dramas of which supplied the armatures about which other civic interactions arrayed themselves. Wagner saw his own proposed festival in a similar light. In a famous section of *The Four Elements of Architecture* (1851), Semper wrote, attacking entire schools of architectural history as he did so, that architectural opinion overlooks the more general and less dubious influence that the carpet, in its capacity as a *wall*, as a vertical means of protection, had on the evolution of certain architectural forms. Thus I seem to stand without the support of a single authority when I assert that the carpet wall plays the most important role in the general history of art.[40]

The craft of the weaver is crucial to this understanding of the wall. Semper's *wandbereiter*, or wall-fitter, is an architectural figure specifically disinterested in the load-bearing, stereotomic arts of the mason. The weaving-in of symbolic meaning to the carpet-wall, as the portable and theatrical device most essential to the social dramas of early civic association, at least as guessed at by Semper, appears also as one of the ideological accoutrements of the proscenial political dramas of 1848 and the role of his barricade in them. It figures too, in a further feature that Semper brought to the auditorium at Bayreuth. His organization of the applied decoration of the auditorium includes the disposal of the Corinthian columns that dress the supports for its visually and acoustically uninterrupted space. These decorations are arranged to perspectival effect. The handling of the scale and intervals between the columns as they approach the stage, especially at the inner and outer proscenium, produces for the audience not only an impression of giganticism for the onstage agonists but a sense that the drama unfolds at some mythic distance. Ironically perhaps, in Semper's articulation of the walls of Bayreuth, Wagner's perspective-worrying, musically chromatic fluxes are architecturally dramatized and augmented by the naturata of Semper's diagrammatically achieved deceit.

A further acoustic detail was perhaps for some imperceptible. It adds to and refigures the urban political themes of this perspectivalism. The second act of *Die Meistersinger* starts with intrigue, is interrupted by lessons on probity, and ends in riot. During this, Sachs the craftsman, and Sixtus Beckmesser, the wretched bureaucrat, musically and spatially signify the differing authenticity of their cultural presences. The narrative here is littered with thwartings and interruptions of different kinds. On two specific occasions, once when he sees Eva Pogner and Walther von Stolzing attempt an elopement on foot, and once when he sees Beckmesser approach the Pogner residence to woo Eva at the window, Sachs sets to exercising his cobbler's skills, singing, while loudly beating shoes on a last. On both occasions he successfully prevents an undesirable course of action. On both occasions he takes the opportunity to lecture. On both occasions, with that instructional and learned hammer blow, as loud as any pistol shot or firework, he signifies an index of cultural presence.

Convention has it that Hans Sachs, as performed, will hit a shoe. The blow will be augmented in the strings, but it is a shoe that he hits. There will be acoustic if not exactly musical authenticity. The syncopations produced by this beating are central to the form of both the music and the narrative over the next twenty minutes. This is so, even though, by the end of the second act, the meaning of the blow has migrated from a preserving celebration of worthy national cultural virtues of handcraft to a punishment beating meted out to Sixtus Beckmesser for his lousy songwriting and his importune ardency. This is a comic beating, intended to

6.15 Interior of the auditorium at Bayreuth. Note the columnation, and especially the perspective-modulating inner and outer proscenial arches. Photograph: Jörg Schulze. © Bayreuther Festspiele GmbH.

signify the utter civic humiliation of Beckmesser and the murder of his cultural reputation. On the other hand, Beckmesser, performed as an at first seriously menacing figure by Heinrich Pflanzl in the Knappertsbusch recording, is required to mug, to engage in eye-rolling ludicrousness, and, importantly, to *mime*. The lute, which Beckmesser carries to accompany his lambasted song to Eva but which, for the stage, he is clearly *not* playing, is ventriloquized by the orchestra's harp. For Knappertsbusch's reading at Bayreuth, with the conventional seating of the orchestra, rather than that adopted by Karajan, the harp may have been as much as twenty feet away from Pflanzl, deep below the stage and at the back of the steep, downwardly raked orchestra pit.

This distinction of instrumental immediacies for Sachs and Beckmesser has its meanings for Wagner's allegory. Sachs is lent cultural authority with that indexical hammerblow. The presence and authentication signaled by it licenses his habit of advisory interference. Also, the onomatopoeia of the "Jerum, Jerum" with which he accompanies the rhythm speaks of *yesterday*, of a latinized, mediaeval sense of legitimating tradition. It is because of this authority that he can see and admire the spontaneous, unreflective aristocratic prodigy of Walther von Stolzing, for instance. Beckmesser, who as Marker in the singing competition in act 1, had earlier indicated, graphically, in chalk, the manifold failings of Stolzing's songwriting, comes himself to be judged by the authority of Sachs's hammer. At all points after his interruption of the elopement of Eva and Walther, the tools of Sachs, the poet-shoemaker, describe the vernacular locus of aesthetic probity. In this scene,

6.16 The raked coiffure of Heinrich Pflanzl singing Sixtus Beckmesser at Bayreuth, 1952. Photograph: Liselotte Strelow. Courtesy © Bayreuther Festspiele GmbH.

Beckmesser, the bureaucrat, in reaching to tradition for his own authority, is rendered increasingly culturally irrelevant and disconnected by each error nailed by Sachs. Here, for one version of his cartoonery, it is thought that Wagner may have publicly crucified his critic Hanslick.

Crucially, in the Knappertsbusch recording of *Die Meistersinger* in 1952, neither of the interruptive hammer blows delivered by Otto Edelmann in his performance of Sachs is heard. The microphones at these narratively vital points are, to borrow from another language, overexposed. All that is heard is anacrustic—a diminuendo, the trailing resonance of the blow, as it bounces off the walls and ceiling of the auditorium, and as it is absorbed by the bodies of the audience, who, of course, hear the blow fully, absorbing its meaning utterly, exhausting it. The radio audience to this performance, at the time, may not have read allegorical significance into this nonpresence of the blow. The longer *stain* on the memory of the blow may have fulfilled and rendered this nonpresence imperceptible. But they may have heard it, nevertheless.

These two hammer blows, interrupting Eva and Walther at one point, Beckmesser at another, lauding one, humiliating another, are key moments in the text of *Die Meistersinger*. They represent instants where the audience is invited to sympathetically cathect with the narratives of German cultural presence. Postwar, in 1952, in the context of the de-Nazification of Bayreuth, and in the further context of Konrad Adenauer's inclination to follow the interests of the Allied administration in Germany, and to rule out the reunification, despite Stalin's apparent offer, these two missing blows may nevertheless come to speak of something; precisely in the nature of their nonpresence. Despite all the other rustling reality effects of this recording, which may assist in the fabrication of a radio listener's desire for the shareable presence of the audience at Bayreuth, to commune in the unilluminated space of the auditorium, Sachs's correctional interruptions are themselves interrupted. They are technically, rather than intellectually or aesthetically, deleted.

———

The microphones, supplied to capture Knappertsbusch at Bayreuth in 1952, may surrender such cultural-political ambiguities in the representation of the space of the hall. This was an intentionally monophonic recording, which is to say one that relies only on volume and intensity to figure its spatialities, and as such is readable in the manner of Desmond Shawe-Taylor and Edward Sackville-West. There was a particular temporal coherence underwriting it. It is and was known to be a live recording, taken from a single performance.

This is unlike Karajan's early Bayreuth essay, which is collated from several. In this sense, then, the nonpresence of Sachs's blows may be thought anchored by one specific indexical spatiotemporal acceptance.

The case with Karajan's second recording of *Die Meistersinger*, the one made in stereo in Dresden in the early 1970s, is rather different, especially in terms of its *time*. The spatial anchors of this recording are visual—photographic, in fact, and, in the end, fabular. This later recording of the same passages from act 2 involves Theo Adam's interpretation of Sachs. His same hammer blows are heard clearly enough, though almost as distant claps. Beckmesser's part is taken by Welsh bass-baritone Geraint Evans. His then widely recognized abilities in playing the braggart buffoon are exercised to comic effect, even though this expertise is not made available in any visual way, except perhaps where, in the documentation of the recording in the liner notes, it may be mnemonically improvised from the *stain* of Verdi's Falstaff on his photographic portrait.

What is also interesting is the way that Evans's rolled *r*'s and his nasal whine work in this. For a European audience, renewed in its apprehensions of alterity, Evans's Welsh accent goes a great way in relocating the meanings that are caught up in the narrative of the "Jew in the Thornbush," the tale often taken as the representation of the culturally dislocated character of Beckmesser's bureaucrat.[41] New to Wagnerian roles, lacking somewhat the culturally nuanced German diction of Heinrich Pflanzl in 1952, Evan's comic voice is shaped as

6.17 Geraint Evans (left) and Theo Adam (right) during rehearsals for the EMI recording of *Die Meistersinger* in 1970–71. Photographs © Siegfried Lauterwasser.

much by the heroic precociousness of the chapel and rugby field as by the vocal requirements of Verdi and the special sympathies of his own nationalism. The subtleties of Evans's braggadocio are heard clearly in his arguments with Sachs, as his attempts to woo are interrupted. Most vividly, they are heard as he sings the first line of his song to Eva, under the agreed instruction of Sachs, who again will strike every time Beckmesser "unsings" himself.

With the words: "Den tag seh' ich erscheinen / der mir wohl gefallen tut" (The day I see appear / which pleases me well), two things becomes apparent:. first, Evans's Welsh intonations and syntax, and second, the cool resonances of the architectural space of the Lukaskirche with its expectant silence into which Evans sings, and which is brought to light with his warbling of the word *erscheinen* (appear).

Sachs strikes, and what is clear from the evidential resonance of the blow is that it was made in a different space. It may be that it was sounded in one of the booths that can be seen in the photographs in the accompanying liner notes—perhaps. From these photographs it is possible to see the paraphernalia of the recording and the appearance of the church itself: the booths, the placement of those telling, though largely distracting Neumann microphones, the distribution of the musical forces, the place of Karajan, his hair, the arrangement of the recording desk—all of which, though superficially convincing, in fact relate very little concerning the aural spatialities of the finished product. The sound of Sachs's blow is interrupted, on its way to Evans and elsewhere, both by a microphone and by the walls of a booth. In fact, the sound never reaches either Evans or even, significantly, Theo Adam's Sachs. It is not he, in this performance, who strikes the blow. The blow's sound simply doesn't figure in the larger dislimning delineation of the Lukaskirche occasioned by Evans's voice. This effect is something that may be achieved by recording from a booth with a limiting resonance, by using a microphone with a limited range, or more simply, by just opening a window.

Variously, the expectations of synchronic presence caught up in the spatial musical conventions of *Die Meistersinger*, as sharpened by the poetic effects of the hammer blow, become unhinged here. In part, these disappointments are shared in the Knappertsbusch and in the later Karajan recordings. All the tiny, convincing reality effects of the Knappertsbusch recording were (are) available to the audience of the recording and its radio audience. But they were not necessarily available to those present at Bayreuth on July 12, 1952.

However, the temporal stake is different for Karajan at the Lukaskirche. Each of these recordings depends on separate regimes of microtemporality for its achievement of architecture. The monophonic Knappertsbusch recording depends on the differences in the short times taken for sounds to reach a microphone, and as

6.18 Interior of the Lukaskirche in Dresden, ca. 1970–71, at the time of the EMI stereo recording of *Die Meistersinger*. Note the microphones, especially perhaps the Neumann M-50. Photographs: © Siegfried Lauterwasser.

they bounce off the walls of Bayreuth auditorium and are absorbed by the bodies of the audience there. They are determined, however, in their production of monophonic space solely by the respective *volumes* of direct sound and resonant sound.

The Karajan recording relies for its difference on another, much smaller psychoacoustic temporal differential, one that is fundamentally exploited by stereo techniques. This, the microtemporality of the Haas or *precedent* effect, governs the smallest unit of time by which it is possible to trigonometrically determine the direction from which a sound comes.

A third microtemporality is an interruption. It involves a temporal contraction effected by microphonically taking the sound made in one part of the hall straight to the mixing desk, without allowing it to figure in the more general aggregate sonority of the space. It is this microphony, in seeking to address the theatrical interests of a specifically phonographic, rather than a radiophonic audience, that shapes the architectural separation of Karajan's Sachs from his hammer. This phonographic audience was one habituated to aesthetic and ethical differences between the historically developing spatialities detailed here: the undecidable monophonies of the Knappertsbusch recording, the inventiveness of what was heard by Sackville-West and Shawe-Taylor, the spatial totalitarianism of the RCA and Mercury minimal pair arrays, and the more juicily pluralized and discontinuous spaces conjured by John Culshaw and Georg Solti for Decca. And there is a great irony too. Wagner's opera presents an ancient, preindustrial image of noble artistic expression where authentic invention is grounded in a firmly settled social order, as signified by the metrical irregularity of Sachs's hammer. Yet these recordings, especially von Karajan's, rely for their poetic spatiality on the tiniest, most refined subdivisions of aural-architectural time. This time is brought about by the spatial effect of musical metrical time, which is itself but a subset of the forms of industrial time that help hold in place a modern class structure: getting laborers to work on time, fixing wages and rates, and allowing in that process for the relative safety of capitalist speculation. In Dresden, just on the other side of the Iron Curtain, Herbert von Karajan's microphones were involved in a profound social and cultural wager.

———

To recall Semper's *wandbereiter*, the use of microphones at the Lukaskirche and the blending together of what they hear did not attempt a linear perspective. Rather they played a role in the reticulation of an aural architectural fabric. On one hand, this fabric is something that is much akin to the kinds of figural function of the line in *trapping* an audience that were described by Jacques Lacan in the relevant sections of *The Four Fundamentals of Psychoanalysis*: those that deal with the psycho-poetic relationships between line and light. In this regard Lacan's writing might serve as a theory of microphony.

On the other hand, such an indication of fabric and line might parallel a rather earlier text. Winston Churchill's Sinews of Peace address was given in 1946.[42] It was here that Churchill revised the meaning of the term "Iron Curtain." For all Churchill's talk of the strong parent races of Europe in this monumental speech, for all his focus on the essentially western European character of Italian Communism and its difficulties with Stalin, his was a speech profoundly figured by notions of reticulation. On one hand there is a reticulation of the interests of American and British imperial endeavors. Here, the international significance of the Commonwealth countries appears shot through by a thread of cultural consistency: the administrative concerns of British imperialism. On the other hand, there is a reticulation of the interests of the Soviet bloc countries, with their different order of consistency. With this precursory statement, the idea of the line drawn through Germany in 1952 and which divided Bavaria, separating Dresden from Munich and Bayreuth, comes to epiphenomenally represent an interface of these reticulates. Here a nexus of political conflicts, negotiations, proposals, and refutations supplies a set of contesting predicates about the meaning of Germany, and attends to those figures of Germany represented by the agonists of *Die Meistersinger*.

Representing, as it were, musical vanishing points in the political contexts of the meaning of Germany, at the respective moments of their recording both of the orders of acoustic nonpresence and temporal theft found in these productions of *Die Meistersinger*, announces moments that may reflect on those circumstances for different types of phonographic audience. These recordings each sponsor a redirection to some or another symbolic "elsewhere" of the hitherto indisputable presences of key actors in Wagner's florid romantic comedy. This is part of an ongoing historical drift away from a mimetic aural-spatial appeal to an idealized concert hall experience, as effected and represented over time by Karajan's use of the possibilities of stereo sound manipulation.

These recordings are part of the textural contradictions found in a representation of Germany to an international audience of Germany (or, rather, of many Germanies possible at that time). Each in its own ways collaterally comes to meaningful terms with the divisions articulated by a line drawn through Berlin in August

1961. The suggestion that there is something to be made of the acoustic articulation of this dramatically divisive urban line, in light of its similarities to the hairlines of Karajan and Tschumi, is not so silly.

4 Recrimination

In their temporal differences, in their different cultural political moments, these two recordings act to produce what might be regarded as a recognizable, ambiguous, acoustic constellation of the meaning for the Berlin Wall. They hold it up, they permeate it, they show it to be a relationship rather than a division. Somehow, they bring together the Stalin Note of the summer of 1952, and Willi Brandt's signing of the Transit Agreement and the Basic Agreement, in 1972. Although it may be pleasing to imagine a poetic-cognitive relationship between the line drawn through Berlin, and the fluxing meanings of the political fringes of Karajan, Knappertsbusch, Elvis, and others, in fact, as Gottfried Semper's view of the wall-fitting architect implies, it is not the *line* that is entirely at stake here.

––––––

"An obedient daughter speaks only when asked," observes Eva Pogner, at the start of act 2 of *Die Meistersinger*. "How wise, how good," nods her father. This account of a suggestive conversation between Tschumi and Wagner has been conducted exclusively through male figures. But the nostalgically dreamed *étatist* politics, the nonpresences articulated here (which are really neither absences or lapses), also point to an otherwise engenderable set of intrigues.

No matter how delighted or irritated a critical audience may become with the performances of here Lisa Della Casa, there Elizabeth Schwarzkopf or Hilde Gueden, or elsewhere Helen Donath, the character of Eva Pogner appears superficially as little more than a squabbled-over token of desirably distracting if somewhat *gemütlich* youth.[43] The words critically expended on Herbert von Karajan's controversial choice of singers for the roles of Beckmesser and Eva at the Lukaskirche all return to two themes. First: the anecdotally relayed reassurances of Karajan to his protégés that he, like Wagner, was incapable of aesthetic error. And second: the indexically appropriate years of Helen Donath. Within the constraints of the tiring erotic landscape of *Die Meistersinger*, in her blonde wig, Eva is pushed biddably about. This is in much the same ways that may be expected of mild-mannered female operatic characters. A love interest, she is not Carmen. Eva is at all points drawn as subject to her desire to please. She wants to please Veit Pogner, Hans Sachs, Walther von Stolzing, Sixtus Beckmesser (more politely, it has to be said), and even, at an amicable remove, her maid's lover, David, the excitably inexpert apprentice to Hans Sachs.

That wish to please is itself a reticulated subjective space. It carries with it the resolution of contradicting forces: the various modes of address required of Eva by her differently vain male leads and the impression of Eva's own desirable subjective integrity. This reticulated subjective space is repeated elsewhere in a Wagner's *oeuvre*. The Cosima Wagner of her published diaries was a bitterly nostalgic, recriminatory individual. In her plangent refrains about how things used to be, especially when she detailed the endlessly redrawn trajectory of her husband's relationship with Gottfried Semper, as well as her own pious and merciless cruelty toward Semper, we find fear, especially of abandonment. Along with everything else, she recorded her husband's dreams. One of these dreams came at the time of Wagner's attempts to wangle the plans for the Festspielhaus out of Semper. Her recounting of it perhaps figures a pathological desire to support and to please. She wrote:

R. had a curious dream: Semper with a plaster mask, annoyed at being recognised, the Wesendoncks, I suddenly vanished, anxious searching, till suddenly my voice as clear as silver but frightened is heard calling "Richard." He cannot reply, which surprises me, but at last, struggling to wake up, he calls out, "Here I am."

The narrated status of this dream evokes very much of one of the key moments where Sigmund Freud developed a category of *wahn*, for his own purposes. This enormously broad and encryptable term was one with special domestic and aesthetic significance for the Wagners: their home *Wahnfried*. In his close reading of the fictionalized dream at the heart of Wilhelm Jensen's short story *Gradiva*, Sigmund Freud proposed the meaning of *wahn* as a species of waking delusion. Cosima's recording of Richard's dreams, as well as what she may have inadvertently contained in them by her surprise at them, is clearly available to some form of literary interpretation. Cosima's act of recording may say something of her anxieties about her husband's desires for a relationship with Semper—one that, by this point, Semper was unlikely to agree to. It may say something of a worry that if the civic role of Wagner's wife disappears, so too does she. Her account may parallel the sublimated behavior of Eva, and further parallel even that of Della Casa, Schwarzkopf, Gueden, and Donath to Knappertsbusch or to Karajan. The whole may even evince a certain marital iciness toward Richard and his itinerant libido. These are all indexical functions that demand a policeable, symptomatic connection between historically archived psychic materials and their meanings.

The superficies of Eva hide the phallic characteristics of the role. Eva's ambiguities lie, in part, in her plural availability to the libidinal requirements of each of the male supports in Wagner's drama. While she may be reduced to a prize, she cannot represent the same prize to each. Though minor in terms of singing time, the role of Eva is a crucial one. It is Eva, after all, who sings into hierarchical relationship the various characterizations of civic probity assumed by the drama. She does so in a way that Beckmesser, in his wretchedly instrumental ambition, is utterly unable to. The very slightness of the role revises many of the aural-architectural issues raised by the performance of *Die Meistersinger* here. At those moments when the staging of *Die Meistersinger* comes forward, Eva represents a formational appeal to its always various audiences, especially those international audiences who are keen perhaps to see how the subjectively fraught issue of German nationhood might respond to its most recent political tests. The almost transparent figure of Eva presents in a suturing capacity. In providing the promise of a singular and pertinent image to each, she definitively embraces all and refuses the superiority of any one possibility. This of course depends on how Eva is sung. The phallic and intimidating insistence of Elizabeth Schwarzkopf tells a different tale from the more reticent affective qualities of Lisa Della Casa's voice, for instance. Nevertheless, it is in Eva that the thoroughly *sinthomic* aspect of *Die Meistersinger*, as offered by the recording of the spaces of its performances, announces itself. And, it is in this sense, for the manifold, microphonic, aural-architectural conditions of *Die Meistersinger*, that Eva represents the key figure of an *acoustic* drama.

Postscript

We should return to a view of another culture of the 1970s, to which this essay has continually alluded. In September 1979, Siouxsie and the Banshees and the Cure played at Sophia Gardens in Cardiff, since destroyed and rebuilt. I saw them. This was at a time before the Wagnerian, romantic darkness of the term "Goth" had come to figure significantly in the reputations of these or other post-punk bands.

The concert took place also only a matter of days since reports that guitarist John McKay and drummer Kenny Morris of the Banshees had made unexplained departures. After the Cure's supporting set, the Banshees took the stage. Budgie, a long-standing friend of the band, filled Kenny Morris's place. Robert Smith of the Cure took McKay's role. Holding his then trademark instrument, a Vox Teardrop electric guitar, Smith read for chord changes and progressions from a sheet of paper on a chair, upon which he occasionally rested his foot. At once authentic and inauthentic, his delirious eye-rolling distraction was occasionally compromised by a need to concentrate on the more mundane business of anticipating middle eights. There was something of Beckmesser here. Susan Dallion, in singing Siouxsie Sioux, addressed the crowd, and gave them license to beat Kenny Morris and John MacKay on sight, for their abandonment of her. I was so happy to be there.

It was here that I first *saw* a plane of variously styled hair as a legitimately transient architectural feature of concert halls. For all the pink kilts, bondage gear, and plastic bike jackets, for all the Dr. Martens shoes, zipped suede winkle-pickers, and scruffied blazers, for all the rest of the burbling sartorial vocabulary that lay beneath this plane, it was the complexity of an adrenalized flux of hair, lit fragmentedly from the stage, that figured the cultural complexity of that moment. And, whether it be raked and rowed as in the auditorium at Bayreuth, or back-combed and bobbing in Sophia Gardens in the late 1970s, that coiffeured plane remains a viably Tschumian feature of the aural-architectural event, for precisely as long as the music plays.

[...] Particular Failings

Before the puzzle was boxed and readdressed

To the puzzle shop in the mid-Sixties

Something tells me that one piece contrived

To stay in the boy's pocket. How do I know?

I know because so many later puzzles

Had missing pieces—Maggie Teyte's high notes

Gone at the war's end, end of the vogue for collies

A house torn down. And hadn't Mademoiselle

Kept back her pitiable bit of truth as well?

—James Merrill, "Lost in Translation"

Incompletion

Despite the evidence and his hopes for being understood, I guess we'll never really know what, when, or, most likely, who James Merrill's "puzzle shop in the mid-Sixties" was. Equally, it is not possible to be precise about the meaning of Maggie Teyte's lost notes, at least not in the way Merrill presents their condition here in "Lost in Translation." Haunted by a room, the further burden of this, probably his best-known poem, is that of the disappointment of the completeness of the puzzle. Or, rather, the puzzle resolved otherwise, joined together elliptically, and in a way that doesn't require all of its given pieces. Here Teyte's notes are given as an element in the structure of an enigma, certainly. But, with Merrill, her voice is also a handsome domestic appurtenance. Her lost notes, her lost purchase on the world, point to the textures of another's urban place and urban time, not hers; to the suggestion of the sensibility of another cenacle founded in a poetic of lost and longed for security, again not hers. What Merrill describes as gone from Teyte's voice, an absence he takes to be as concrete an object in a psychological landscape as a destroyed home, describes a presence in the interests of a poet, even if one so abstractly wealthy and leisured.

So, here in the context of one of the most widely read twentieth-century American poems, the presence of the nonpresence of sounds, their relationship to both tangible as well as nontangible architectures, and the tenuously named forms of society that exist in that space are given as familiar things.

For myself, I think Maggie Teyte made a bit of a racket with her voice, and not just in her twilight years. It's not just that technical inadequacies appear when her singing is compared to other, sometimes less well-known figures. Other things emerge in the fabric of her voice—overtones of regional accent at a particular time and class presumption. There seems to be a pursed imperiousness to the glottal grip she takes on some of the songs she championed, and which shape those presumptions. I have wondered if James Merrill could have even heard, let alone had an opinion on those details. In any case, none of these things, undesirable as they may seem, will prevent me from listening with interest to her recordings, or to those of Gwen Cately for example, or Irene Joachim—singers whose voices sit at a similar kind of remove from universalized ideals of female vocality, whose voices grew less competent in conventional terms over time, yet ever more texturally complicated and musically innovative in their apparent failings. I like listening to Maggie Teyte's voice because I don't like the sound of it.

And that is not such a perverse or capricious thing to say. Infatuations often appear in the form of self-criticism. Part of me doesn't like her voice. Another part of me enjoys the sometimes rather cruelly revealing sport of exploring why. But, more importantly, an observation like this, one that is capable of setting oneself aside from oneself, marks the beginning of the formulation of a question, the phrasing of which can proceed in the discernible manner we saw raised by Gabriel Marcel earlier. It isn't always possible to know what one is listening for until one is engaged in—actively caught up in—listening. One thing is clear: for Merrill, Teyte's failing voice doesn't have musical meaning here. Listening euphemistically, it is easy to make absences into presences in the way James Merrill does, and to compose, with others, one's own form of acentric interest in things.

Changing the light on objects, translating them from one discursive form of their existence to another, often prompts a sense of augmentation in their range of possible meanings. Sound objects become architectural; architectural devices become shyly musicological, enhance, expand, extend their poetic selves in the process of migration. The same process of movement, however, invariably leads to a sense of that profit in meaning coming at some cost. There is some inkling of inaccessible dimensions that are privately retained by the original matrix that fostered the object's integrity and significance, and which are sacrificed on its entry to another, perhaps more public orbit. Yet, that assertively evident lack is itself also a locus of speculation, of wondering, of further meaning-making—one that is regulated only by the capacities of the audience who witnesses the shift in register and by what that audience is then prepared to make of that lack as a creatively exploitable surplus.

Two points have become quite striking in the course of this book's unfolding. First, it is possible to see that each chapter is marked by some account of a cherishable insufficiency in sound, as if sound might only become architectural by dint of its inadequacy, and as if, further, the lure and seduction of inhabitable acoustic space is found in its need for supplementation, social, aesthetic, or otherwise. What I mean to say by this is that, in their capacities to connect and cathect concrete with psychological spatialities, the structure of aurality and the structure of desire have something in common. Whether it has been in the way the sound of architectural water is seen to remark on physical and emotional crises, or the way the inability of radios to apprehend and broadcast the entire dynamic range of a piece of music came to shape a suburban sensibility, or how it was that a simple absence of vocabulary in a second-hand shop pointed up discomforts concerning different kinds of transactions in places of learning—wherever it has been, what we have seen here is the spatial quality of sound expressed in terms of relationships to a condition of their fallibility, even their delinquency.

The second point is that the spatial sounds we have encountered here are also particular. They are not generic or representative. One sound can't stand in for another like it. The sounds here, in their complexity, cannot play a part in any general theory of modern sound and modern space. The theorems and devices that have been used to animate the problematics that these sound-complexes might articulate are not free-standing or portable forms. They are, rather, bespoke and specific. The perception itself that has recognized the objects in question is quite overtly a part of the object that the perception animates—animates, that is, not ventriloquizes.

What has also been evident in each chapter of this book is the tenuous and uncertain structure of the sounds and spaces that it has announced. I have made an attempt to allow the reader to assemble his or her own subject of conversation in each chapter. To that end, I have tried to relate, for instance, the singing of Walt Disney's dwarves to Mies van der Rohe's love of figured stone, and to do that to a certain extent elliptically. In that way, a view of labor, for instance, isn't the only thing that holds them together. Viewed from the point of view of sound, architecture comes to extend and endure in unfamiliar ways. It is able to take on different types of burden. Recognizing that is part of an effort to think of architecture not in terms of its load-bearing capability, or its shelter-giving form, or the socially instrumental lucidity of its visual grammar, but rather to think of the correct connective tissue of aural architecture to be […] the ellipsis, questionable.

He is dead now, so maybe it doesn't matter if I say that James Merrill seemed to inhabit the ellipse. Already liberated by fabulous wealth, he looked for different orders of social and intellectual liberty by inhabiting with others the cherished fallibility, even implausibility he could find in assembled things. This is perhaps more important than it sounds. Histories of art and architecture and music and so forth become surly when objective cultural historical facts, their constitution and the given relationships between them, are challenged subjectively. Merrill's poem points to an idea that the conscious manufacture of objects and the terms of one's interest in them are crucial to genuinely speculative modes of cultural interpretation. Moreover, the act itself has more general, collateral effects on the person engaged in it. In his much-loved, much-loathed essay, "Ideology and the Ideological State Apparatuses," Louis Althusser worked toward a notion of liberating abilities of aural architecture with a concept of interpellation. This is a sense of the configuration of space through a sensation of being called or hailed by one's given cultural environment. It is marked by a radical lack of doubt, in those moments of being addressed by a culture, that the culture means *you*. Not someone else; you. Only you. Althusser unpicked the places where ideology and affects merge together to make this a political difficulty and suggested that were one able to respond to the space-making effects of interpellation with the answer "Do you mean me?" there would have already been a step toward liberation.

Imagining Louis Althusser conversing amiably in a room with James Merrill isn't such a crazy idea; it might be one to add to the canon of twentieth century aural-architectural objects. What is clear, though, is that Merrill had already surpassed him. In the situation that Merrill authored for his self poetically, in "Lost in Translation," the only sensible reply to the question "Do you mean me?" is a singular "Yes." From there it might be possible to see the aural rearrangement of architecture and its responsibilities as a step toward the authoring of cultural liberty.

Notes

Introduction: Architecture, Aurality?

1. For his work in this area, see Leo Beranek, *Concert Halls and Opera Houses: Music, Acoustics, and Architecture* (New York: Wiley, 1962). For a continuation of similar cultural historical study, see M. Forsyth, *Buildings for Music: The Architect, the Musician, and the Listener from the Seventeenth Century to the Present* (Cambridge: Cambridge University Press, 1985).

2. For Helmholz's best-known work on the psychophysics of musical perception, see his *On the Sensations of Tone as a Physiological Basis for the Theory of Music* (London: Longmans, Green, 1885).

3. See Nouvel's discussion of some aspects of the design of the Culture and Congress Centre at Lucerne, in his "The Exact Representation of a Will," in *Luzern Konzertsaal* (Berlin: Birkhauser, 1999), 73–89.

4. *The Record of Singing*, EMI, RLS 724, 1977.

5. *Berlin: Die Staatsoper Unter der Linden (1919–1945)*, Merian Musik-Dokumentation, BASF 98 22177-6, 1976.

6. Mary Ann Doane, *The Voice in Cinema: The Articulation of Body and Space*, Yale French Studies (New Haven: Yale University Press, 1980), 33–50.

7. See A. M. Pescatello, *Charles Seeger: A Life in American Music* (Pittsburgh: University of Pittsburgh Press, 1992).

8. See Beatriz Colomina, "Battlelines," in *The Architect: Reconstructing Her Practice*, ed. F. Hughes (Cambridge, MA: MIT Press, 1996).

1   Ballerinas Are Always Hungry

1. G. Marcel, *Metaphysical Journal* (London: Rockliffe, 1952), 121.

2. Quoted in *Performance Artists Talking in the Eighties*, ed. L. M. Montano (Berkeley: University of California Press, 2000), 382.

3. Principally the old Spread Eagle Hotel, on Corporation Street, in central Manchester, where Moss maintained an office.

4. See O. Newman, *Defensible Space: Crime Prevention through Urban Design* (New York: Macmillan, 1972). For key accounts of the ambitions and rhetoric of the new urbanist movement, see *Charter of the New Urbanism*, ed. M. Lecesse and K. McCormick (New York: McGraw Hill, 1999), and P. Katz, *The New Urbanism: Toward an Architecture of Community* (New York: McGraw Hill, 1994).

5. See, e.g., M. Dolan, *The American Porch* (New York: Lyons Press, 2002).

6. This same diary was to be published as his *Metaphysical Journal* ten years later.

7. Marcel, *Metaphysical Journal*, 149.

8. For discussions of the term *modernism* for Christian thought in the twentieth century, see A. Lilley, *The Programme of Modernism* (London: Fisher Unwin, 1908), and L. P. Jacks, *The Revolt against Mechanism* (London: George Allen & Unwin, 1933). One of the key forums for the pursuit of this argument was Jacks's own Unitarian-inspired periodical, the *Hibbert Journal.* This was also one of the places where the social and doctrinal implications of theological arguments as diverse as those advanced by Alfred Loisy, Henri Bergson, and Jacques Maritain were introduced to an English-reading audience.

9. A *charrette* is a professionalized and rapid-thinking investigative forum preferred by new urbanist social inquiry. It was designed as a means of getting municipal officers, planners, community representatives, and business leaders into conclave, so that any issues that may hinder a development may be identified swiftly and overcome.

10. The implications here of the term *non-authored* derive from Jean-Luc Nancy's usage in his critique of the determining first-person position. See J.-L. Nancy, "Of Being Singular Plural," in his *Being Singular Plural* (Stanford: Stanford University Press, 2000), 15.

11. E. Underhill, *The Mystic Way: The Role of Mysticism in Christian Life* (1913; London: Ariel Press, 1992), 189.

12. S. Maharaj, "Black Art's Autre-Biography," In *Run through the Jungle: Selected Writings by Eddie Chambers*, ed. G. Tawadros (London: InIVA, 1999), 4, 8.

13. In many senses, the arguments of Schumacher's 1972 book *Small Is Beautiful* follow in a complex tradition of Christian social thought articulated in Leo XIII's papal encyclical *De Rerum Novarum* (1891). This document was relied upon to a great extent by the theorists and advocates of distributism in the 1930s, including the likes of G. K. Chesterton, Hilaire Belloc, A. J. Penty, and others. It should be noted that distributism, in its own special understanding of the individual's rights to land ownership, for example, has been rightly regarded as a thoroughly reactionary doctrine, and has influenced much of the retreatism of third way thinking in the United States. In this sense, distributism is the awkward background to the ideas offered by Marcel and Antin.

14. E. Jacobsen, "Receiving Community: The Church and the Future of the New Urbanist Movement," *Markets and Morality* 6, no. 1 (spring 2003): 67.

15. Newman, *Defensible Space*, 186.

16. See H. Rosenau, *The Ideal City in Its Architectural Evolution* (London: Routledge & Kegan Paul, 1959).

2 Omissibility

1. L. Goldmann, *The Hidden God: A Study of Tragic Vision in the Pensees of Pascal and the Tragedies of Racine* (London: Routledge & Kegan Paul, 1964), 287.

2. A. Schoenberg, *Fundamental of Musical Composition* (London: Faber & Faber, 1967). He worked on the text from 1937 to 1948.

3. J. Betjeman, "Love Is Dead," in *First and Last Loves* (London: John Murray, 1952), 11.

4. It is worth remembering that *The Pioneers of the Modern Movement* initially reached a relatively small audience. It wasn't until Henry-Russell Hitchcock et al. exercised their particular skills for visual historical argument and worked to supplement the nature and volume of photographic illustration of the text for the Museum of Modern Art's edition in 1949 that the book started to develop its more widespread and popular appeal—and this in the context of Pevsner's own increasingly influential position in the world of architectural letters.

5. R. Williams, *Television: Technology and Cultural Form* (London: Fontana, 1974).

6. Here I allude to the two opening sentences of Channan Willner's essay "The Two-Length Bar Revisited: Handel and the Hemiola," in *Göttinger Händel-Beiträge* (vol. 4, 1991). A generous musicologist, Willner is one of a few who have contributed to the recent development and exploration of the inordinately complex area of the function and perceptibility of effects of metrical dissonance, such as the hemiola, in the Western orchestral

and choral repertoire. Others who have also contributed significantly to the discussion include Floyd K. Grave, David Schulenberg, Julian Kahlberg, Tilden Russell, Justin London, Richard Cohn, Harald Kreb, and Vincent Corrigan.

7. F. K. Grave, "Metrical Dissonance in Haydn," *Journal of Musicology* 13, no. 2 (1995): 168–202.

8. D. Schulenberg, "Commentary on Channan Willner, 'More on Handel and the Hemiola,'" *Music Theory Online* 2, no. 5 (1996).

9. The most complete and reflective account of Panofsky's methodology remains his own, *Studies in Iconology* (Oxford: Oxford University Press, 1939).

10. E. Panofsky, *Studies in Iconology* (Oxford: Oxford University Press, 1939), 7.

11. Pevsner was an important historian of the Baroque in the German tradition of his mentors Wilhelm Pinder and August Schmarsow. His doctoral work was on the Baroque of southern Germany. For a discussion of the implications of these antecedents, especially the National Socialist cultural associations of Pinder, see M. Halbertsma, "Nikolaus Pevsner and the End of a Tradition: The Legacy of Wilhelm Pinder," *Apollo* 137, no. 372 (1993): 107–110.

12. For such disagreements regarding the ways that novel historical methodologies were taken to represent characteristic national sensibilities, see the text of a pair of radio talks given by each in 1952. N. Pevsner, "An Un-English Activity—I: Reflections on Not Teaching Art History," *Listener* 47, no. 1235 (October 30, 1952): 715–716; and E. Waterhouse, "An Un-English Activity—II: Art as a 'Piece of History,'" *Listener* 47, no. 1236 (November 6, 1952): 761–762.

13. N. Pevsner, *The Pioneers of the Modern Movement* (London: Faber & Faber, 1936), 206.

14. R. Stone, "Turtlewax, or; 'There Ain't No Flies on the Lamb of God,'" *Parallax* 1, no. 2 (1996).

15. H.-R. Hitchcock, "Some Problems in the Interpretation of Modern Architecture," *Journal of the Association of Architectural Historians* (1942): 29.

16. T. A. Adorno, *Introduction to the Sociology of Music* (New York: Seabury, 1967), 264.

17. D. F. Tovey, "Symphony in E Flat Major, No. 5, Op. 72" (1935), in *Essays in Musical Analysis: II Symphonies* (Oxford: Oxford University Press, 1936).

3 Reveals: Glass Houses, Stones

1. L. Trotsky, *The History of the Russian Revolution* (London: Pluto Press, 1977), 1155.

2. W. Benjamin, "Surrealism," *Reflections* (New York: Harcourt Brace, 1978), 226.

3. R. Caillois, *The Writing of Stones*, trans. Barbara Bray (1970; Charlottesville: University of Virginia Press, 1985).

4. M. Cacciari, "The Glass Chain," in *Architecture and Nihilism: On the Philosophy of Modern Architecture* (New Haven: Yale University Press, 1993), 190.

5. R. Caillois, "Surrealism as a World of Signs" (1978), trans. Claudine Frank and Camille Naish, in *The Edge of Surrealism*, ed. F. Frank (Durham, NC: Duke University Press, 2003), 327.

6. I take this gesture toward oracularity from Jacques Lacan's theorization of the *sinthome* and the creatively laboring, self-criticizing analysand. See *The Seminar of Jacques Lacan: On a Discourse That Might Not Be a Semblance, 1971*, translated by Cormac Gallagher, published privately.

7. For example, Cage's friendship with composer Virgil Thomson seems to have suffered because of disagreements over what should and shouldn't be included in his biography of Thomson. For some notes concerning Cage's tribulations over a biography of Thomson, see D. Revill, *Roaring Silence: John Cage, A Life* (London: Bloomsbury, 1992).

8. The East River Housing Cooperative (initially known as the Corlear's Hook Project) was established by the United Housing Federation and the International Ladies Garment Workers Union (ILGWU). It was the first development in New York City to benefit from slum clearance funds under Roosevelt's Federal Housing Act of 1934.

9. Russian emigré Morris Sigman was founder of the International Workers of the World organization and was president of the ILGWU in 1923. Benjamin Schlesinger was manager of New York's *Jewish Daily Forward* and also president of the ILGWU. Morris Hillquit was a practicing lawyer who helped found the American Socialist Party. Henryk Ehrlich and Viktor Alter were leaders of the Jewish socialist organization in Poland, the Bund.

10. Prominent labor leader David Dubinsky was president of the ILGWU in the 1930s when the struggles between Communists and non-Communists brought the organization to the brink of collapse. He oversaw the rise to influence of the organization through its relationships with the instruments of Roosevelt's New Deal policies. Louis "Lepke" Buchalter, on the other hand, was head of Murder Inc. Among the many things Buchalter was accused of was using the threat of strike action on the part of the garment workers' unions in order to hold employers to ransom for his personal gain.

11. See Jane Jacob's classic polemic against the practice of the deleting the organization and details or architecturally reified cultural formations from New York's urban theater, *The Death and Life of Great American Cities* (London: Jonathan Cape, 1962).

12. For one of several discussions of memorial and amnesiac functions in the incoherent history of New York's urban planning policy, see M. Page, *The Creative Destruction of Manhattan, 1900–1940* (Chicago: University of Chicago Press, 1999).

13. Revill, *Roaring Silence*, 95.

14. Cardew in a conversation on *Music Now* (BBC 2, January 2, 1970). See K. Spence, "Television," *Musical Times*, March 1970, 303.

15. See the argument pursued in section 2 ("Indeterminacy") of the lecture Cage gave at Darmstadt in 1958, "Composition as Process," in *Silence: Lectures and Writings by John Cage* (Middletown, CT: Wesleyan University Press, 1961).

16. Cage, *Silence: Lectures and Writings*, 37.

17. It should be remembered that, in attempting to avoid the specifically figured window on the musical piece afforded by the character of the virtuoso performer, Christian Wolff, among others, sometimes went out of his way compositionally to frustrate David Tudor's schooled habit of fully working out a score before a performance, in order to limit Tudor's agency as mediator between score and sound.

18. See N. Pevsner, "The Architectural Setting of Jane Austen's Novels," *Journal of the Warburg and Courtauld Institutes* 31, (1968): 404–422.

19. See C. Cardew, "John Cage: Ghost or Monster," *Listener* 87, no. 2249 (May 4, 1972): 597–598; reprinted in C. Cardew, *Stockhausen Serves Imperialism and Other Essays* (London: Latimer New Dimensions, 1974).

20. For discussions of this particularly concerning Cardew's work, see B. Dennis, "Cardew's *The Great Learning*," *Musical Times*, November 1971, 1066–1068; and V. Anderson, "Chinese Characters and Experimental Structure in Cornelius Cardew's *The Great Learning*," *Journal of Experimental Music Studies*, http://www.users.waitrose.com/~chobbs/jems.html.

21. C. Cardew, *Stockhausen Serves Imperialism*, 94.

22. Ibid., 98.

23. It is worth adding a footnote to Cardew's political development. Cardew died under mysterious circumstances, a hit-and-run incident. There has been much speculation about this. In 1982 a memorial concert was staged at the Queen Elizabeth Hall in London. This included performances of passages from his *Treatise*, as well as *Octet*, *Paragraph 1* of *The Great Learning*, and some of his later worker's songs. Unsurprisingly, the concert attracted the various milieus of experimental music performers in the UK, many of whom were on stage that night. A recording of the event was published by Impetus Records in 1985. On the sleeve notes the following disclaimer appears: "The Cornelius Cardew Foundation would like to make it clear

that later in his life Cardew rejected Maoism." Many institutions have attempted to play down the overtness of the way Cardew pursued his politics, which were themselves in any case criticizable on their own stated grounds. Regarding Mao, in one sense Cardew did make a repudiation. He repudiated the position of Lin-Piao, someone who died in equally mysterious circumstances. And, during the Sino-Albanian split, he made a claim for the hardline political stances adopted by Enver Hoxha, and for Albania's increased openness to other European states.

24. See E. Prevost, *No Sound Is Innocent: AMM and the Practice of Self-Invention* (Matching Tye, Essex: Copola, 1995).

25. C. Cardew, *Scratch Music* (London: Latimer New Dimensions, 1972), S, T.

26. Ibid., C.

27. C. Cardew, "Toward an Ethic of Improvisation," in *Treatise Handbook* (London: Peters, 1971), xvii.

28. For an aggressive characterization of Mies van der Rohe's presence in Germany at this time, see E. S. Hochman, *Architects of Fortune: Mies van der Rohe and the Third Reich* (New York: Wiedenfeld & Nicholson, 1989).

29. Sometime after completing this discussion of Mies and Cage, I came across Branden Joseph's important article "The Architecture of Silence" (*October* 81 [1997]: 66–74). Although the present chapter can't claim to stand on Joseph's observations on the phenomenal aspects of Miesian glass (and in ways differs from them), I should admit to a retrospective debt.

30. L. Mies van der Rohe, "What Would Concrete, What Would Steel Be without Mirror Glass" (March 13, 1933), in Fritz Neumeyer, *The Artless Word* (Cambridge, MA: MIT Press, 1991), 314.

31. See W. Tegethoff, *Mies van der Rohe: The Villas and Country Houses* (Cambridge, MA: MIT Press, 1985).

32. "Mies Speaks," *Architectural Review* (December 1968): 452. Transcript of an interview for RIAS in 1966.

33. Caillois, "Surrealism as a World of Signs," 75.

34. Ibid., 101.

4 Splash

1. R. M. Rilke, *Sonnets to Orpheus* (1922; Berkeley: University of California Press, 1960), 113.

2. G. Louganis and E. Marcus, *Breaking the Surface* (New York: Sourcebooks, 2006), 75.

3. I should thank Phil Rudd for the way that he has lent interpretative significance to this aspect of the practice of architectural photography, in his doctoral research into the work of Julius Schulman and other photographers.

4.   F. L. Wright, "The Architectural Forum," in *Frank Lloyd Wright: Collected Writings*, vol. 3, ed. B. Pfeiffer (New York: Rizzoli, 1994), 280.

5.   C. Kingsley, *The Waterbabies: A Fairytale for a Land-Baby* (London: Macmillan, 1886), 4.

6.   R. Durgnat, "Movie Eye," *Architectural Review* 137 (March 1965): 817.

7.   As a fragmentary pun on the representation of the sea and its identifications, the phrase "deline the mare" is taken from a couplet: "Wild sea money. Dominic Deasy kens them a'. *Won't you come to Sandymount, / Madeline the Mare?* Rhythm begins, you see. I hear. A catalectic tetrameter of iambs march ing. No, a gallop: *deline the mare.*" J. Joyce, *Ulysses* (Oxford: Oxford University Press, 2011), 37.

8.   See F. L. Wright, "The Man Who Was Gertrude," in *Frank Lloyd Wright: Collected Writings*, vol. 3, 260–261.

9.   For a discussion of the way the term *iteration* may imply an ever increasing degree of play in the repeated use of a term, see Jacques Derrida's essay "Structure, Sign, and Play in the Discourse of the Human Sciences," in his *Writing and Difference* (Chicago: Chicago University Press, 1978). In other critical spheres, and to more influential effect, the term has also been severally exercised by Judith Butler. See, e.g., her *Bodies That Matter: On the Discursive Limits of "Sex"* (New York: Routledge, 1993).

10.   F. L. Wright, *The Living City* (New York: Mentor, 1952), 97.

11.   Health and Safety Commission, *Managing Health and Safety in Swimming Pools* (Wetherby: Sport England, 2003), 64.

12.   Ben Johnson has said of his *The Unattended Moment* that when it was hung near the swimming pool of a private house, when the viewer looked at the reflection of the painting in the real pool and the depiction of reflections in the painted pool, the sounds of the water heightened the illusion of the space portrayed.

13.   F. L. Wright, *The Living City* (New York: Mentor, 1963), 97.

14.   See P. Collins, *Changing Ideals in Modern Architecture, 1750–1950* (Montreal: McGill-Queens University Press, 1965).

15.   For Mark Wigley's exercise of alimentary figures in the deconstruction of architectural gestures, see his *The Architecture of Deconstruction: Derrida's Haunt* (Cambridge, MA: MIT Press, 1995).

16.   In 1988, Jacques Benveniste claimed that it was possible that water molecules have the ability to "remember" the signatures of other molecules. Derived from digital study of molecular forms, his findings were applauded by homeopaths and derided by others in the scientific community. Before his death in 2004, Benveniste proposed that it might be possible to reduce these memories to a digital aural format, leading to a possibility (or fantasy) of online cures.

17. For Le Corbusier, see his *La Ville Radieuse* (Paris: Vincent, Fréal & Cie, 1933), 148. For Jencks, see his *The Architecture of the Jumping Universe* (New York: Academy Editions, 1997). Despite the claims to a significantly postmodern perspective in Jencks's manifold histories, there are remarkable historiographic similarities between, for example, his overt claims and those assumptions rather brilliantly buried in the complex account of the ontogenesis of modern architecture offered by Nikolaus Pevsner in his *The Pioneers of the Modern Movement* (1936).

18. See G. Simondon, *On the Mode of Existence of Technical Objects* (Paris: Aubier Editions, 1958).

19. Advertisement for Swimclear Pool Filter Aids, *Swimming Pool Review* 13, no. 2 (June 1972): 291.

20. P. Morton-Shand, *Architectural Review* (March 1935): 32.

21. This purview has been criticized by writers as diverse as Beatriz Colomina, Michel de Certeau, and Leslie Martin. See, e.g., Leslie Martin, *Urban Space and Structures* (Cambridge: Cambridge University Press, 1972); B. Colomina, "Battle Lines," in *The Architect: Reconstructing Her Practice*, ed. F. Hughes (Cambridge, MA: MIT Press, 1996); and, as indicated at the beginning of this essay, M. de Certeau, "Walking in the City," in *The Practice of Everyday Life* (Berkeley: University of California Press, 1984).

22. B. Jones, "Public Swimming Baths," *Architectural Review* (May 1967): 342. It is worth noting that in making his comments on public pools, Jones passed over other notable pools, for example, the one built for its employees by Shell on the South Bank of the Thames in 1957, and the pool complex commissioned by Harmerson at Holborn in lieu of council ground rent in 1960. Both of these have rather social meanings in terms of leisure that are rather different from Dryburgh's Empire Pool.

23. J. Richards, "The Pioneer Health Centre," *Architectural Review* (1938).

24. Anon., *Observer*, December 21, 1935, 4.

25. Anon., "Ecology in Peckham," *St. Bartholomew's Hospital Journal* (September 1936): 229–233.

26. David Goodfinch, Chief Assistant Architect (Health) for Leeds City Council, was a significant polemicist for such building types in requirements to the demands of the National Health Act (1946). Leeds, like other towns in northern England, was famously tardy in producing plans for action concerning slum redevelopment during this period, and that such icons of socialist housing as the flats at Quarry Hill in Leeds were built at all is a tribute to a demanding resilience on the part of municipal figures like Goodfinch. See *Architectural Journal* (July 3, 1947): 12.

27. Ibid.

28. J. Lewis, "The Peckham Health Centre: An Inquiry into the Nature of Living," *Bulletin: Society for the Social History of Medicine* (June–December 1982): 30.

29. J. Lewis and B. Brookes, "The Peckham Health Centre, 'PEP,' and the Concept of General Health Practice During the 1930s and 1940s," *Medical History* 27 (1983): 45.

30. F. Safford, "Health Centres for Preventative Medicine, Recreation and Education," *Architectural Record* (February 1938): 6661. Safford also noted that some of his American medical colleagues recognized little demand for preventative medicine. He thought that, doubtless, "the present campaign against syphilis is giving this demand new impetus."

31. T. Benton, "The Pioneer Health Centre," *Architectural Design* 49, nos. 10–11 (1979): 56–59.

32. From the TV documentary *Mustn't Grumble*, BBC Television, 1989.

33. I. Pearse and L. H. Crocker, *The Peckham Experiment: The Study of the Living Structure of Society* (London: Allen & Unwin, 1943), 91.

34. Benton, "The Pioneer Health Care Centre."

35. G. S. Williamson, *Annual Report of the Peckham Health Centre 1936* (London: Pioneer Health Centre, 1936), 3.

36. Cited in Benton, "The Pioneer Health Care Centre."

37. I. Pearse, *The Quality of Life* (Edinburgh: Scottish Academic Press, 1979), 79.

38. I. Pearse, *The Peckham Experiment* (London: Allen & Unwin, 1943), 231.

39. Moreover, Horace's Epistle to Numicius (Book 1, Epistle 6) seems to provide a neat mythological identity for Ned Merrill.

40. Interview with Barbara Walters, *20/20*, ABC, February 24, 1995.

41. Louganis and Marcus, *Breaking the Surface*, 185.

42. F. Rich, "A Bigger Splash," *New York Times*, March 12, 1995.

5   Blush

1. E.-L. Boullée, "Architecture: Essay on Art," in *Boullée and Visionary Architecture*, ed. H. Rosenau (London: Academy Editions, 1976), 95.

2. S. Freud, "On Fetishism" (1927), in *The Standard Edition of the Complete Psychological Works of Sigmund Freud*, vol. 21 (London: Hogarth Press, 1961), 150.

3. M. Klein, "Infantile Anxiety Situations Reflected in a Work of Art and the Creative Impulse," in *The Selected Melanie Klein*, ed. J. Mitchell (London: Hogarth Press, 1986), 84–95.

4. Julia Kristeva has recently insisted on Klein's subscription to the "cult of the narrative." She has, however, also noted in Klein's broader attentions, a perception of the artwork as "the initial, perhaps optimal way of caring for people." See J. Kristeva, *Melanie Klein* (New York: Columbia University Press, 2001), 104, 186.

5. St. Bonaventure, "The Life of St. Francis," in *The Little Flowers of St. Francis* (London: Dent, 1910), 358.

6. A. Smithson and P. Smithson, *Without Rhetoric: Architectural Aesthetic, 1955–72* (London: Latimer New Dimensions, 1973), 75.

7. Ibid., 76.

8. A. Brendel, *Musical Thoughts and Afterthoughts* (London: Robson Books: 1976), 147.

6 Waxing on Walls

1. D. Hussey, review of BBC broadcasts of performances from Bayreuth in 1952, in *Listener* 47, no. 1235 (October 30, 1952): 737–738.

2. B. Tschumi, *Theoretical Projects: The Manhattan Transcripts* (New York: St. Martin's Press, 1981), 44.

3. For his architectural appropriation of Kafka, see B. Tschumi, "Burrow/Earth," *Domus* 610 (October 1980): 4.

4. R. Barthes, "The Romans in Film" (1954), in *Mythologies* (New York: Hill & Wang, 1972), 26–27.

5. Neumann produced a series of directional and omniphonic microphones in the immediate postwar years that were to transform the procedures of music recording. The U-47 (1949) was the first of the high-caliber switchable microphones that allowed for different patterns to be adopted with the same instrument. The introduction of a remote switching capacity in 1951, along with new capsule technology that allowed for a particular smoothness at each end of the dynamic range, paved the way for the mythical M-50 microphone with its four pattern options and especially effective omniphony. This latter type of microphone is favored for placing above large orchestras to obtain the overall architectural image.

6. B. Tschumi, "Index of Architecture: Themes from the Manhattan Transcripts," in *Questions of Space* (London: AA Files, 1983), 107.

7. Volkseigener Betrieb (VEB) Deutsche Schallplatten was established as a state-owned enterprise in 1953.

8. R. Osborne, *Herbert von Karajan* (London: Chatto & Windus, 1998), 587. Osborne notes that a recording of an emotional speech that Karajan made to the crew and performers at the end of these recording sessions, in which he mentioned the shine of the Staatskapelle, was mysteriously lost. He also notes the presence of a mysterious member of the Stasi.

9. G. Gould, "The Prospects of Recording," *High Fidelity Magazine* 16, no. 4 (April 1966): 46–63.

10. tIt should be noted that the spatializing effects caused by a reduction in the use of bass was a recurrent theme in such journals as the *Gramophone* during the mid-1930s. See, e.g., the discussion by P. Wilson, "Diffusion," *Gramophone* 14, no. 161 (October 1936): 217.

11.   For an introduction to the style of these writers, see E. Sackville-West and D. Shawe Taylor, *The Record Guide* (London: Collins, 1955). Compare with the critical language of Warren de Motte in *The Long Playing Record Guide* (New York: Dell, 1955), who focuses on interpretation at the expense of any other criteria.

12.   T. W. Adorno, *In Search of Wagner* (1954; London: NLB, 1981), 15.

13.   A. Aber, "Tradition and Revolution at Bayreuth," *Musical Times* 92 (1951): 262.

14.   The first extended electric recordings made of *Die Meistersinger* at Bayreuth were those of the performances conducted by Karl Muck in 1925.

15.   Irene, the daughter of Fritz Busch's brother, the noted violinist Adolf, married Serkin in 1935. They had been close for years. The Busches remained in voluntary exile from 1933, returning only in 1951.

16.   It is worth noting that in an illustrated index to *The Manhattan Transcripts*, Bernard Tschumi also included a number of images from Paul Wegener's 1920 film *The Golem*. The images of Prague in the sets for this film, which were designed by Hans Poelzig, bear a striking architectural resemblance to some of Reissinger's images of Nuremberg for the Bayreuth Festival.

17.   Aside from the overtly interested critics of and apologists for Wagner, commentary on *Die Meistersinger* falls into two broad schools. These are exemplified on one hand by the scholarly and on occasion engagingly defensive journal *Wagner*, and, on the other, the highly critical stance offered by books such as Nicholas Vazsonyi's edited collection *Wagner's Meistersinger: Performance, History, Representation* (Rochester: Rochester University Press, 2003).

18.   The use of the term *stain* here is derived from Jacques Lacan's usage in the passages on the thing (*l'achose*) in his seminar of Wednesday, March 10, 1971. It refers to a concept of the projection of the historical coherence of the thing onto an object, and represents an important locus of the functioning of the realm of the Lacanian symbolic. See J. Lacan, *The Seminar of Jacques Lacan, Book XVIII: On a Discourse That Might Not Be a Semblance* (1971), private translation from unedited transcript (Cormac Gallagher, 1971), V1–V20.

19.   I should thank Elizabeth Lebas for an anecdotal account of the way her mother used the first half of a recording of Bach's *St. Matthew Passion* as both metronome and chronograph to her daily house-cleaning routines in the early 1960s.

20.   Alan Blumlein is credited with the invention of stereo recording techniques in the 1930s. These could not be commercially exploited at the time, as no means of transferring the recordings to a viable support could be devised until the later 1950s. Even then, the problem of agreeing on industry standards was not fully resolved until 1958. Blumlein's use of two microphones placed an appropriate distance apart, augmented by a third microphone to one side, converts spatial position into a microtemporal, directional difference between the

time taken for a sound to reach one and then another microphone. His method apprehended the essential concept on which later techniques were based. Problems with microphone spacing were perennial for a time. Put the microphones too close together, i.e., below the threshold of perceivable directional-temporal difference, and there is no stereo staging; too far apart and a "hole" appears in the middle of the image. There were also other more deliciously comic effects of spatial distortion and incoherence, such as an apparent giganticism of the players, frequently remarked upon at the time. Arthur Wilkinson contrived the Decca Tree stereo microphone array in 1954. Arthur Haddy and Arthur Wilkinson, both of whom were involved in military uses of sound during the Second World War (especially sonar), were able to develop such a technology for commercial and aesthetic use in the subsequent period. The Decca Tree used M50 Neumann microphones widely spaced, with a third placed in front. The standardized French Radio method used a pair of directional microphones placed much closer together, though aligned in different directions. A variation on this, usually credited to Swiss engineer Jürg Jecklin, involves an acoustically opaque baffle placed between the microphones, to prevent each microphone picking up what the other does. The radically reduced format of a minimal coincident pair of microphones is the format that, though rather arid sounding, is preferred by many audiophiles. It was used for the prized Mercury Living Presence and RCA Living Stereo recordings. Each of these microphone arrays produces different spatial effects. Each has sponsored complex subjective critical vocabularies to define its aesthetic qualities, and each has contributed to an enormously sophisticated aural grammar.

21.   Throughout the summer of 1952, political relations between the Allies and the Soviets in discussions over the future of Germany were to a great extent forged over what exactly may have been meant by Stalin's offer to reunify Germany, and for it to become a nonaligned, self-governing nation. On the one hand, it has been speculated that Stalin's offer held some legitimacy, especially in terms of supplying a means to provide a national barrier between the Soviets and the Allied Forces. On the other hand, suspicions about Stalin's true intentions in opening a negotiation, and what might ensue from that, mean that several historians have seen the lost opportunity of reunification in 1952, and Konrad Adenauer's apparent complicity in this, as something of a myth. In any case, it seems unlikely that this issue will be resolved without the emergence of further archival materials. For a preliminary introduction to the historiographical themes of this extensive debate, see R. van Dijk, "The 1952 Stalin Note Debate: Myth or Missed Opportunity for Reunification," *Cold War International History Project*, Working Paper No. 14, May 1996.

22.   J. Ranciere, "The Shoemaker and the Knight," in *The Philosopher and His Poor* (Durham, NC: Duke University Press, 2004), 57–70.

23. C. Debussy, "*Parsifal* et le Société des Grandes Editions de France," *Gil Blas*, April 6, 1903; reprinted in *Wagner* 3, no. 2 (May 1982): 45–47. Despite his reference to German conductors, Nikisch, as Debussy pointed out, was in fact Hungarian, and Cortot, French.

24. R. Rolland, "A Note on Siegfried and Wagner," in *Musicians of Today* (London: Henry Holt, 1915), 221.

25. F. Nixon, "The Changing Climate of Opera," *Tempo* (June 1952): 25.

26. E. Bloch, *Essays on the Philosophy of Music* (Cambridge: Cambridge University Press, 1985), 223.

27. See R. Wittkower, *Architectural Principles in the Age of Humanism* (London: Studies of the University of London, 1949).

28. The case of Redon is instructive. Though he produced a series of speculative images of Brünnhilde and Parsifal, his parallels with Wagner run in other directions more convincingly. These are perhaps best exemplified in the efflorescences of scintillating color used in the flower paintings, which perhaps repeat the chromatic devices employed by Wagner, and the layering of plural symbolic spaces.

29. A. Lorenz, "On the Musical Structure of Die Meistersinger," *Wagner* 3 (1981): 9–13. It is also worth noting that with this mode of analysis of the formal apprehension of aesthetic emotion, Lorenz prefigured later modernist approaches to the cultural embodiment of meaning (Clement Greenberg, for example, or Michael Fried) and was working in a tradition that included his contemporary Clive Bell.

30. T. Adorno, *In Search of Wagner* (1952; London: NLB, 1981), 66.

31. T. Adorno, "Wagner's Relevance for Today," trans. S. Gillespie, in *Essays on Music*, ed. R. Leppert (Berkeley: University of California Press, 2002), 592.

32. For two exemplary discussions of the social and philosophical implications of the architectural line, see M. Nesbit, "The Language of Industry," in *The Definitively Unfinished Marcel Duchamp*, ed. T. de Duve (Cambridge, MA: MIT Press, 1991), and C. Ingraham, *The Burden of Linearity* (New Haven: Yale University Press, 1998).

33. For a discussion of Knappertsbusch's alleged role in the exile of Mann, see H. R. Vaget, "The Wagner Celebrations of 1933 and the National Excommunication of Thomas Mann," *Wagner* 16, no. 2 (May 1995): 51–61.

34. See E. Preetorius, "Richard Wagner: Stage Picture and Vision," *Wagner* 12, no. 2 (May 1991): 75–86.

35. This recording of the performance given on July 12, 1952, initially released as an LP set by Melodram (MEL 522) in 1981 as part of the 12 Jahre neu Bayreuth (1951–1962) series, appears to be taken from the archives of Bavarian Radio. It has been rereleased on CD, in variously restored forms by a number of publishers.

For reference, listen to the version released by Hans Knappertsbusch-Gesellschaft, Munich, and Golden Melodram, GM.1.0003.1997.

36. With reference to the character of this preferred mode of fantastic attention, I should isolate the following two remarks. (1) "The second shock came with the little phrase on the back cover: 'Registrazione dal vivo'—i.e., live recording. Nobody had warned me about that! I hate live opera recordings, with all their stage noises destroying every opportunity to let your own fantasy roam free, if not obscuring the music itself." This comment, from reviewer Martien Philipse, is taken from a four-star review (posted on Amazon.com, August 17, 2003) of a much-admired recording of *Die Meistersinger* conducted by Rafael Kubelík in 1967. Philipse went on to say: "Fortunately, however, this Meistersinger was recorded in the Herkulessaal in Munich, during a concertante performance that is, and if there was an audience (which I doubt), it is inaudible." (2) "The advantage of the recording is that you can imagine her on any stage, anywhere in the world." This remark was made by Andrew MacGregor, presenter of BBC Radio's Saturday morning CD review program, in providing a review of a recording of excerpts from operas by Richard Wagner and Richard Strauss, sung by Deborah Voight (BBC Radio 3, April 11, 2004). Voight is the remarkable soprano who had, only a few weeks previously, been dismissed from a production of Strauss's *Ariadne auf Naxos* at Covent Garden, London, because of her weight. Here, in articulating a sheerly aural pornography, the absence of audience-generated atmosphere and the use of multimiking techniques in simulating architectural spaces (and the construal of other kinds of spatial fiction) all make for the plausibility of a critical framework that is quite reliant on the isolation of performance.

37. The first of the digital recordings of Wagner's music included the full Ring cycle recorded by Marek Janowski for Eurodisc and VEB Deutsche Schallplatten, again at the Dresden Lukaskirche venue, in Dresden and again with the Staatskapelle, in the early 1980s. They are almost terrifyingly clear as recordings, revealing all sorts of musical details that, according to Wagner's aesthetic intentions, should be lost in each other as a blended, overall timbre.

38. Compare the ambiences of Karajan's *Die Meistersinger* recorded with Columbia (33CX1025) with Knappertsbusch's recording of *Parsifal* with Decca from the same year (LXT 2651–56).

39. For one account, see Ernest Newman, *The Life of Richard Wagner* (London: Cassell, 1933).

40. G. Semper, *Four Principles of Architecture and Other Writings* (Cambridge: Cambridge University Press, 1989), 103.

41. For this referential locus of interpretations of Beckmesser as a figuration of anti-Semitism, see J. Grimm and W. Grimm, "The Jew in the Thornbush," *Wagner* 18, no. 2 (May 1997): 91–96.

42. Churchill delivered his Sinews of Peace address at Westminster College, Fulton, Missouri, on March 5, 1946. The term "Iron Curtain" was first used by Joseph Goebbels in February 1945 in his role as Minister of Enlightenment and Propaganda.

43. Lisa Della Casa was the Swiss soprano who played the role of Eva in the Knappertsbusch recording of the 1952 Bayreuth performance. This was her first outing at Bayreuth. She despised the intrigue and scheming that went on, and never returned. Elizabeth Schwarzkopf sang Eva in the Karajan recording at Bayreuth the previous year. Romantically involved with EMI producer Walter Legge, it is perhaps fortuitous that she should have declined the role the year after, in the knowledge that the Decca team were at Bayreuth to record as many performances as they could. Hilde Gueden sang Eva for the much-loved Decca studio recording conducted by Knappertsbusch, and which included an iconic performance by Paul Schoffler as Sachs. Helen Donath sang for Karajan's recording at the Lukaskirche in Dresden in the 1970s.

44. M. Gregor-Dellin, *Cosima Wagner's Diaries*, vol. 2: *1878–1883* (New York: Harcourt Brace Jovanovich, 1978), 177.

Index

Teyte, Maggie, 215–217
Tietjen, Heinz, 184
Tovey, Donald Francis, 59
Tschumi, Bernard, 169–179, 182, 189–190, 192, 210, 213

Vaughan Williams, Ralph, 55, 166

Wagner, Cosima, 211
Wagner, Richard, 102, 165, 169–170, 178–179, 181–184, 186–198, 200–204, 209–212
Wagner, Wieland, 183, 194
Wagner, Winifred, 194
Williams, Owen, 95, 117–118, 120–122, 124
Williamson, George Scott, 122, 124, 128, 138
Wintersgill, H. H., 45–46
Wittkower, Rudolf, 5, 64, 78
Wolff, Christian, 3, 67–69, 73, 76, 80, 84, 87
Worm, Diether Gerhardt, 179
Wright, Frank Lloyd, 22, 85, 95, 100–101, 105–106, 108, 110, 116, 143